"Just as loving parents thoughtfully rese[...] the impact it has on their children, church leaders must do the same for those Jesus entrusts us with. Too often we jump on the latest trends and whatever seems most attractive on the surface, without much thoughtful discernment. *Analog Church* is a wake-up call and asks us some tough, much-needed questions—whether our rush into the use of whatever new technology is available is helping or hurting people's understanding of God, worship, church, and themselves. In a digitally saturated world, where new generations are bombarded and immersed in the digital, we need to press into analog all the more. *Analog Church* shows us how."

Dan Kimball, director of the ReGeneration Project, author of *They Like Jesus but Not the Church*

"Sometimes the best books about the future involve the ones that start with a look backward. In this very important work, Jay reminds us of God's vision for the church as the plumb line for how we view and leverage technology. In making digital the servant of analog we are moving in the right direction. Reversing the two leads us to something fundamentally different than the deep journey God has called all of us to. The church was always meant to be waiting for us when everything else failed to live up to our deep longing for transcendence. This book is the map to that."

Nancy Ortberg, CEO of Transforming the Bay with Christ, author of *Looking for God*

"Jay Kim is a theological wizard. His writing is sharp without being cutting, pastoral while also prophetic, disruptive but not divisive. In other words, *Analog Church* doesn't just make for good content; it makes for good humans who are ready to trade relevancy for transformation. I can already think of a handful of leaders I want to pass it along to in holy passive aggressiveness or, better yet, love. On my list of things to do after reading it: 'To gather when the world scatters. To slow down when the world speeds up. To commune when the world critiques.'"

Erin S. Lane, author of *Lessons in Belonging from a Church-Going Commitment Phobe*

"Pastors and church planters today face a bewildering variety of options and opinions about how to 'do church' in our contemporary, digital age. It's an age-old dilemma: When do our efforts to 'adapt' to a new cultural setting end up compromising what the Jesus movement has to offer the world? In this book Jay Kim offers a timely and poignant set of reflections that no ministry leader can afford to ignore."

Tim Mackie, cofounder of The Bible Project

"We are clearly sitting within a technological and digital revolution—but a revolution against what? And to where? In this book, my friend Jay Kim serves us as a true pastor, showing us that this revolution is unparalleled in its spiritual implications. After reading this book, I have a much clearer understanding of how technology has shaped the church and how we can change. With an impressive bibliography, thoughtful exegesis of Scripture, and terrific prose, Kim shows us how the digital revolution requires an analog response—and why God's church is the essential respondent."

Chris Nye, pastor and author of *Less of More: Pursuing Spiritual Abundance in a World of Never Enough*

"We're at a point in history where churches are investing considerable amounts of time and money into the digital age. Jay carefully critiques this booming movement with eyes set on a direction far less attractive and all too necessary—detaching more from digital technology and stepping back into patient communion with God and one another. *Analog Church* is not another call to gather the masses and burn it all down but a compelling and, I believe, prophetic invitation to reorient our values to reflect the spiritually enriching practices of generations past. Jay paints us a wonderful picture of life ahead, if we are willing to adjust, where we are faithfully connected to the digital age without being controlled by it."

Zach Bolen, songwriter, producer, and frontman of Citizens

"At a time when so many are despairing over the declining attendance and lack of engagement in our local churches, I cannot wait to give this book to usher in some hope! Jay Y. Kim describes how we can lean into what the church is uniquely poised to provide—transformation, community, and shared moments of wonder and awe. I closed the book grateful for a sense that, with God's help, we can do this!"

Nancy Beach, leadership coach with the Slingshot Group, author of *Gifted to Lead: The Art of Leading as a Woman in the Church*

"In *Analog Church*, Jay Kim rings a bell. He sounds an alarm warning us of the potential dangers inherent in our increasingly disengaged digital age. But he also sounds an invitation—like a distantly familiar dinner bell—calling us back to the transcendent presence of God and the warmth of deeply rich communal life in the kingdom. I resonate deeply with both these sentiments, and I believe you will too."

Manuel Luz, author of *Honest Worship* and *Imagine That*

"It's a grave miscalculation for the church today to think relevance depends on the ability to keep up with the pace, gloss, and hype of our technological world. Our frenetic, fidgety age does not need a frenetic, fidgety church. Our Insta-perfect, polished age does not need a photoshopped, inauthentic church. Our tech-weary world does not need a tech-obsessed church. Jay Kim's *Analog Church* understands this, presenting a compelling case for the church's most radical act in today's world: not to be a trendy, shape-shifting, chameleonic copycat, but to be a transcendent Christ-centered community whose *difference* from the world is why it makes a difference."

Brett McCracken, senior editor at The Gospel Coalition and author of *Uncomfortable: The Awkward and Essential Challenge of Christian Community*

"In his book *Analog Church*, Jay raises important questions and addresses crucial issues for the church in a digital age. Instead of continuing to adapt and acquiesce, he calls us to come out of hiding from behind our digital walls, to bridge digital divides, and to be human with one another in real time, real space, and real ways. He invites us to move beyond relevance to transcendence. And it is a welcome invitation!"

Ruth Haley Barton, founder of the Transforming Center, author of *Life Together in Christ*

"How do you keep real people and genuine fellowship in a virtual world? Jay Kim wrestles brilliantly with the issues and realities of the march to utilize digital technologies both inside and outside the building. *Analog Church* is an essential resource for churches seeking to use digital technologies without falling prey to the disastrous distortions that come to thoughtless adopters."

Gerry Breshears, professor of theology at Western Seminary, Portland

"In our media-saturated age, what we need more than ever is not *relevance* so much as *transcendence*. In *Analog Church*, Jay Kim asks the right questions about our use of technology as churches, regardless of whether one lands with every conclusion, that can help us move from the digital emphasis on information to the biblical emphasis on transformation, from our preferences to others' presence, and from mere communication to the majesty of communion—together as the flesh-and-blood people of God."

Joshua Ryan Butler, pastor of teaching and direction at Redemption Church, Tempe, author of *The Skeletons in God's Closet*

"With wisdom and grace, Jay Kim urges the church to consider the ramifications of the digital age. Without noticing it, we quickly become content with efficiency over intimacy, convenience over transcendence, and results over transformation. *Analog Church* invites us to slow down, to breathe deeply of the human connectedness that we were designed to experience in our communal search for God. This book isn't an invitation to join a sectarian group forsaking all things modern or digital but instead illuminates a relevant, ancient pathway into the profound beauty and mystery of God. I highly recommend this book!"

Kurt Willems, pastor, author, and podcaster at TheologyCurator.com

"Perhaps now more than ever, the church can offer a radical alternative to a digital world. In *Analog Church*, Jay Kim calls us to a greater sense of self-awareness and reminds the bride that she is drastically different, beautiful, and real. Jay opens our imaginations to see a unique, gritty, personal, and embodied path—the path toward transcendence, not relevance."

Tara Beth Leach, senior pastor of PazNaz, Pasadena, California, and author of *Emboldened*

ANALOG CHURCH

WHY WE NEED REAL PEOPLE, PLACES, AND THINGS IN THE DIGITAL AGE

JAY Y. KIM

Foreword by Scot McKnight

An imprint of InterVarsity Press
Downers Grove, Illinois

InterVarsity Press
P.O. Box 1400, Downers Grove, IL 60515-1426
ivpress.com
email@ivpress.com

InterVarsity Press® is the book-publishing division of InterVarsity Christian Fellowship/ USA®, a movement of students and faculty active on campus at hundreds of universities, colleges, and schools of nursing in the United States of America, and a member movement of the International Fellowship of Evangelical Students. For information about local and regional activities, visit intervarsity.org.

All Scripture quotations, unless otherwise indicated, are taken from The Holy Bible, New International Version®, NIV®. Copyright © 1973, 1978, 1984, 2011 by Biblica, Inc.™ Used by permission of Zondervan. All rights reserved worldwide. www.zondervan.com. The "NIV" and "New International Version" are trademarks registered in the United States Patent and Trademark Office by Biblica, Inc.™

While any stories in this book are true, some names and identifying information may have been changed to protect the privacy of individuals.

"Worship," "Community," and "Scripture" figures are used with permission, courtesy of Jessie Barnes.

Cover design and image composite: David Fassett
Interior design: Jeanna Wiggins
Images: abstract group of people: © dan4 / iStock / Getty Images Plus
      silhouette of hand: © CSA Images / Getty Images
      Bible illustration: © CSA Images / Getty Images
      old paper texture: © Katsumi Murouchi / Moment Collection / Getty Images

ISBN 978-0-8308-4158-5 (print)
ISBN 978-0-8308-4198-1 (digital)

Printed in the United States of America ♾

Library of Congress Cataloging-in-Publication Data
A catalog record for this book is available from the Library of Congress.

P 25 24 23 22 21 20 19 18 17 16 15 14 13 12 11 10 9 8 7 6 5

Y   37   36   35   34   33   32   31   30   29   28   27   26   25   24   23   22

For Jenny, my favorite.

CONTENTS

FOREWORD by Scot McKnight   1

INTRODUCTION | **EDM AND GRANDMA'S CHURCH**
The Relevance of Transcendence   5

1 | **SLOW AND STEADY**
Why Go Analog?   13

PART 1 – WORSHIP

2 | **CAMERAS, COPYCATS, AND CARICATURES**
Worship in the Digital Age   33

3 | **TO ENGAGE AND TO WITNESS**
Analog Worship   56

PART 2 – COMMUNITY

4 | **REBUILDING BABEL**
Community in the Digital Age   81

5 | **A TAX COLLECTOR AND A ZEALOT WALK INTO A CROSSFIT**
Analog Community   101

PART 3 – SCRIPTURE

6 | **JACKPOT!**
Scripture in the Digital Age   133

7 | **HOWTOREADABOOK**
Analog Scripture   152

8 | **THE MEAL AT THE CENTER OF HISTORY**
Communion   172

CONCLUSION | **BLINDED BY THE LIGHT**
Where Do We Go From Here?   181

ACKNOWLEDGMENTS   185

DISCUSSION QUESTIONS   187

NOTES   195

FOREWORD

SCOT McKNIGHT

I WEAR A WRISTWATCH—AN APPLE WATCH, if you care to know—but the face of the watch is an old fashioned set of watch arms: a long one for minutes, a shorter one for hours, and a moving second hand. The terms *analog* and *digital* aren't part of my day-to-day vocabulary, but I suppose you could say my watch is a hybrid of analog and digital. An attempt, perhaps, to have the best of both worlds.

Of course, I'm typing (or keyboarding) this on a laptop computer, which is nothing less than a godsend for any number of reasons. But I remember when I used to write on a piece of paper, text on the top half and footnotes on the bottom half, and then type it into a manuscript on a manual—and later electric—typewriter.

My world is hybridity. The digital parts of my world are convenient (I really have no interest in typing on a manual typewriter anymore) and quick (using digital tools, I once did a bibliography for a book in about seven seconds). These are digital components I've embraced. But I have some limits.

I don't read my Bible digitally (though I do research on biblical texts with Bible software programs). I buy lots of books, almost none of them electronically, and the ebooks I do have I don't use. I want to touch the pages, smell the paper, feel the binding, and underline and mark the pages.

We don't "do" church digitally. If my church went digital, I'd stop going. We meet in a room—more than a hundred of us. We know one another's names, we touch one another, laugh with one another, sip coffee with one another; we see one another's faces and hear the timbre in one another's voices.

There's a theology behind what Jay Kim very helpfully calls *Analog Church*, and it's the incarnation. God became one of us. He became Jewish, in the first century, in a Judea and Galilee ruled by Rome and its underlings. For some people Christianity is digital: God sent a message to us and we pick it up somehow, either believe it or not, and then either live according to it or not. But God didn't send a message. God sent his Son, born of a real woman, married to a real man who had a real job. They all—the whole lot of them—experienced real problems because very few bought their story of a virginal conception. But Jesus grew up and became a real man and found real humans with real bodies to follow him and extend his kingdom mission to the broken and wounded in his part of the world.

If Jesus is God incarnate, then God chose to reveal himself in analog, not digital. You can communicate a message in words and send such a message on paper/papyrus, but you can't see the revelation of God except in that one person— the person who lived, who died on a cross, who was raised up, who ascended, who rules, and who will come again. This is the treat you will find in Jay's *Analog Church*. At that level, *Analog Church* provides an analog ecclesiology that conforms to the incarnation itself.

Church is the same way. We can communicate conveniently and quickly in digital formats—at my church we get an e-newsletter each week, and I like them. But we can't get to know one another apart from embodied realities. One can't "do" church digitally; the important things about church life are all embodied: knowing one another, loving one another, sitting and standing and praying with one another, listening to the sermon and watching the tone of the words and the movement of the body when we sing and walk forward to take communion. These are the things that make a church a church.

And yet the digital age has made some forms of communication, education, and instruction possible in ways previously impossible. We live in hybridity—but hybridity only cuts to the heart and soul if it is rooted in the embodied realities of analog. Kim's book is an important read for those struggling with the inadequacies of our digital age.

EDM AND GRANDMA'S CHURCH

THE RELEVANCE OF TRANSCENDENCE

We're all struggling to say
the same old things in new and different ways.
And so we must praise the new and different ways.

DOBBY GIBSON

MY FRIEND JAKE is an electronic dance music (EDM) artist. I know next to nothing about EDM, but it's a massive subculture, eliciting an almost cult-like allegiance from its fans, who number in the millions globally. I met Jake when he was an insecure but gregarious freshman in high school. It's astounding to see him now, headlining and selling out concert venues around the world.

For those unfamiliar with the world of EDM, imagine . . .

. . . a packed-out nightclub

. . . deafening, high-energy dance music

. . . laser lights cutting through smoke-machine fog, synchronized to the music

. . . large screens onstage, displaying abstract digital visuals

That's what Jake does. He creates spaces like that, night in and night out.

Jake grew up going to church. That's actually how I met him—I was his youth pastor when he was in high school. After graduation, he left the local church for a variety of reasons. On occasion, when he is in town, he still attends church services with his family. He recently told me about one of his experiences there.

The worship gathering was held in a concert venue in a downtown area. As Jake walked into the dimly lit room, he squinted his eyes to see through the smoke-machine fog. As the band began to play, laser lights cut through the haze, synchronized to the beat of the music, which was played at concert-level volume, the massive subwoofers pounding every bass note. A large screen fixed behind the band displayed the lyrics of the songs in dynamic moving text, against a backdrop of abstract digital visuals. After the music, the teaching pastor popped up on screen to deliver the message from a different location, twenty miles away. Describing the experience to me, Jake said, "I didn't feel cool enough to be there. I don't think church should be like that."

Jake's experience at this church was, at least on the surface, uncannily similar to the ones he creates professionally. And yet he felt like he wasn't "cool enough to be there." Jake—an unchurched twenty-something—concluded, "I don't think church should be like that."

Jake is exactly the sort of person most church leaders today are desperate to reach. He's a young millennial (bordering on Generation Z), skeptical of Christianity and cynical about organized religion, but openly searching for hope, meaning, and purpose. What does his recent church experience tell us?

When Jake steps foot in a church, he isn't hoping that it will look, sound, and feel like everything else he's a part of. He isn't searching for *relevance*. He's not concerned about church being cool, hip, or digitally savvy enough for him. In fact, he finds it off-putting to watch a video of a teacher twenty miles away. He's unimpressed that the presentation is showing off the latest and greatest in audio-visual technology. He's on the hunt for something else altogether.

Jake is searching for *transcendence*. He's reluctantly stepping foot in a church in the hope that there might be something there that he can't find anywhere else. Without even knowing it consciously, Jake is looking for something timeless.

And he's not alone. I've heard countless stories of young people who are unimpressed at best, and repulsed at worst, by the modern evangelical church's hyper focus on looking, sounding, and feeling like everything else they see in culture—our red-hot pursuit of relevance. I've heard these stories both firsthand and from the family and friends of young people who've either walked away from the church or were never there to begin with.

Many churches today are working tirelessly to create something that it turns out Jake and others like him aren't looking for. We church leaders want our churches to look and sound and feel just right. We're on the never-ending search for what's new, fresh, and trending. We want people to know that this isn't their grandma's church.

But what if grandma's church actually had some things right? It's easy to criticize and quickly dismiss the pipe organ, choir gowns, and oversized pulpits as antiquated relics of the past, but doing so is an exercise in shallowness and snobbery. What if the steady pace with which they gathered, the intentionality with which they worshiped, the depth of their songs, the richness of their liturgy, the warmth of their conversations, the regular ongoing practice of centering together around the bread and the cup of Communion . . . what if these are all things we still desperately need, maybe even more so now, in the digital age?

Our unchecked pursuit of relevance isn't only affecting the way we gather to worship. It's also changing our understanding of what it means to be a community. As more and more churches push headlong into online spaces, people are being asked not only to communicate but also to commune on digital platforms. The Bible, too, is being affected by the digital age, as we turn the grand narrative of Scripture into a series of easily digestible, bite-sized tidbits for personal encouragement and self-help–style motivation. We're changing the church experience from an extended meal at a dining table into a truncated series of tweets, and we're losing our aptitude for nuance, generosity, and engagement.

One of the most alarming things about all of this is how blind we've become to it. Many church leaders have come to believe that all of this is for the best; we've bought and believed the myth that what's new is always better than what's old. We are guilty of what C.S. Lewis called "chronological snobbery."

Admittedly, I myself have perpetuated this ethos over the years. I've served on staff at a few different churches throughout Silicon Valley for the last decade and a half, including a medium-sized church, a young church plant, and a multisite megachurch. At each, we felt the strong temptation of the digital age—the temptation to pursue relevance at any cost. We found ourselves spending inordinate amounts of time and energy trying to create spaces that looked, sounded, and felt like whatever we thought was most relatable to popular culture at large. Sometimes we chose wisely and sometimes we took things too far.

Ultimately, though, we discovered that any sort of sustained emphasis on relevance invariably led to satisfied Christian consumers who'd found a product they enjoyed but rarely led to anything deeper. The most transformative experiences people were having in our communities, we slowly realized, had nothing to do with the lights, sound, and spectacle. Transformation was happening in much more tactile ways—through personal relationships and the profound simplicity of studying Scripture, praying, and sharing meals together.

This isn't a big-church, small-church issue. Churches of all shapes and sizes are falling headlong into the trap of relevance at all costs, digitizing and technologizing anything and everything they can. Only God knows the truest intentions of the men and women leading the charge. But even when said intentions seem well and good on the surface, the responsibility of leading in the local church demands that we excavate down to the depths of our ambitions and thoughtfully consider the methods we're using to get where we long to go. Yes, as a church leader I want to serve and reach as many people as I can with the gospel. This is true of most church leaders I know. But often, the desire to "serve and reach as many as we can" in the digital age devolves into methods that essentially equate to, "what's the fastest, most efficient way for us to get bigger?" This stands in stark contrast to the sort of growth Jesus himself talked about most (I'll get more into that in the next chapter). As my friend Chris Nye writes, "We cannot abandon the soil for the microwave. . . . Jesus had remarkable patience and pace, which would frustrate the Silicon Valley."[1]

This all came to a head for me one Sunday morning as I was walking to the stage to deliver a sermon I'd worked on for weeks. As I was walking up, the service coordinator reminded me, "Jay, don't forget to look directly into the camera at the back of the room so the campuses feel connected to you."

Look at the camera so the campuses feel connected to you. The camera is the connection. That was the reminder.

In her sobering book *Alone Together*, Sherry Turkle says this: "Digital connections . . . may offer the illusion of companionship without the demands of friendship. Our networked life allows us to hide from each other, even as we are tethered to each other."[2] As I stood on stage, staring into that camera, I felt this tension in real time. I did what was asked. I spoke regularly into that camera, imagining that that small device was the people sitting in rooms all over our city— people I couldn't see, hear, or feel in any visceral, human way.

Something about it felt off. For weeks and months after that experience, even as I did the same thing over and over again, I couldn't shake the sense that there was a better way forward.

In our increasingly digital world, the church has done what she's done countless times before—adapted and acquiesced to contemporary culture. Walk into any church in America today, big or small, and you'll likely be surrounded by digital and technological excellence (or at least the pursuit of it), all for the sake of relevance. This desire to create a church experience that's familiar and appealing to the digitized masses has led to a dangerous miscalculation—that for a church to grow, thrive, serve, and reach its community, it must be on the front edge of the digital and technological age.

But what if there's a different path? What if the church of one hundred and the church of ten thousand are both sitting on untapped mountains of potential?

I believe the answer is to go *analog*. People are hungry for human experiences and the church is perfectly positioned to offer exactly that. In fact, the church is fundamentally designed and intended for this work—to create spaces and opportunities for people from all walks of life to experience true human flourishing, in real time and real space. Unlike anything else in our culture today, the church can invite people to gather in the flesh and to experience the hope that Jesus Christ offers.

Now, you may be saying, "Jay, digital is simply the medium, not the message. Aren't you being a bit nitpicky here?" That's an understandable response, but it denies an alarming reality of what the digital/technological revolution is beginning to reveal—the mediums we use are undeniably intertwined with the messages we share. As the twentieth century philosopher Marshall McLuhan famously said, "The medium is the message."[3] I think he's right. We'll get into this idea in detail in the coming pages.

Since its earliest days, the Christian church has been marked by its invitation to *transcendence*, not *relevance*. This isn't to say that relevance doesn't matter. Our message must be relatable to the everyday lives and circumstances of the people we serve. But the journey there, for followers of Jesus, has always been through unlikely means—what some have called the *upside down kingdom of God*, where the first are last and the last are first, where the rich are poor and the poor are rich, where the outsider experiences belonging, and all who gather can encounter together something totally *other* than iterations of what's already familiar. And in the digital age, one of the most *upside down* things the church can offer is the invitation to be *analog*, to come out of hiding from behind our digital walls, to bridge our technological divides, and to be human with one another in the truest sense—gathering together to be changed and transformed in real time, in real space, in real ways.

ONE

SLOW AND STEADY

WHY GO ANALOG?

Moments of more may leave us with lives of less.

SHERRY TURKLE

N OT LONG AGO, a friend was showing me a couple of the dating apps on his phone. I felt like a dinosaur watching him deftly navigate what seemed like an endless stream of potential dates. Jenny and I started dating in 2003 and have been married since 2009—just about the time when dating websites and apps were in their infancy. So I watched, bewildered, as my friend scrolled quickly through a dizzying variety of profiles, communicating his interest or lack thereof by swiping with ninja-like precision.

I have nothing against online dating. I have several friends who never would have met their spouses if it weren't for technology, and the world is a better place because they're together. The benefits are obvious. What baffled me was my friend's nonchalance and quick trigger. I asked him, "How are you deciding so fast? What if you would've really hit it off with that person? How can you dismiss so many people so quickly?" He looked at me inquisitively and said, "Dude. It's not that serious."

This is our digital world. Even the most important decisions, like the people we choose to enter into meaningful relationships with—maybe even a lifelong commitment—are made with shocking speed. Anything less is considered archaic. And this doesn't just apply to dating: the digital age has affected, and in some cases infected, all spheres of life, including the most vital part of the Christian life: *discipleship*.

Despite our various mission statements, those of us who serve the local church share the same purpose: to introduce people to Jesus Christ and to invite them to follow him with their whole lives. Since the very beginning, this has been an invitation into a lifelong process that involves patiently journeying alongside others. This is what the Bible means when it talks about discipleship—the life of apprenticeship under Jesus, learning and living his ways, being shaped and reshaped into his likeness alongside others. There is no more important endeavor in the Christian life. As Dallas Willard wrote, "The greatest issue facing the world today, with all its heartbreaking needs, is whether those who, by profession or culture, are identified as 'Christians' will become disciples—students, apprentices, practitioners—of Jesus Christ, *steadily* learning from him how to live the life of the Kingdom of the Heavens into every corner of human existence"[1] (emphasis mine).

Willard wrote that discipleship is a process of *steadily* learning how to live the Jesus way. Steady—consistent, unwavering, focused movement in one direction.

The Greek fabulist Aesop gave us the timeless tale of "The Tortoise and the Hare" and with it, the popular idiom, "Slow and steady wins the race." Enmeshed in the speed of the digital age, we've forgotten the moral of this great story. We've forgotten the value—even necessity—of steadiness, slow as it may seem sometimes. Instead, we've given ourselves over to the technological ethos of Silicon Valley, which tells us that everything can always be faster, more efficient, more accessible. And as digital technologies continue to accelerate to lightning speeds, we revel in our dominance and prowess. But as with the hare, our distraction is lulling us to sleep in the all-important journey of *discipleship*.

WHEN VALUES TURN VICIOUS

The digital age's technological advancements boast three major contributions to the improvement of human experience, which in turn have become its undeniable values:

1. *Speed.* We have access to what we want when we want, as quickly as our fingers can type and scroll.

2. *Choices.* We have access to an endless array of options when it comes to just about anything.

3. *Individualism.* Everything, from online profiles to gadgets, is endlessly customizable, allowing us to emphasize our preferences and personalities.

While these contributions have added some comfort and convenience to parts of our lives, the added value is coming at great cost, as our collective desire for and devotion to digital technology becomes increasingly excessive. Particularly in the ways these digital technologies have influenced the church, many of us have gone off the rails. Even good things have dark sides when taken to their extremes. When values aren't held accountable, they turn vicious. Sadly, for so many, that's exactly what's happened. These once positive contributions of the digital age have resulted in our undoing:

The *speed* of the digital age has made us *impatient*.

The *choices* of the digital age have made us *shallow*.

The *individualism* of the digital age has made us *isolated*.

IMPATIENT

According to a research study funded by Microsoft, between 2000 and 2015 the average attention span decreased from twelve seconds to eight seconds.[2] Twelve seconds doesn't seem like a very long attention span to begin with, and to think that, parallel to the rise of the internet, it dropped so significantly in just fifteen years is staggering. The scientific validity of the study has been disputed by some, claiming that attention spans can't be quantified this way because they vary so broadly depending on the specific task at hand.[3] This seems reasonable enough, but what neuroscience is beginning to make clear is that the digital age is indeed rewiring and reshaping us into increasingly *impatient* people.[4]

In his book *The Tech-Wise Family*, Andy Crouch defines digital technology as an "easy everywhere" medium, meaning it's easy to access and it's accessible almost everywhere.[5] Naturally, we've adjusted quickly. Imagine yourself walking into a coffee shop to do some work and discovering that there's no Wi-Fi. Then, imagine you order a drink and sit down to do some internet-less work, only to discover that your MacBook is displaying the spinning pinwheel of death. Now, imagine the overwhelming onset of annoyance you feel. Maybe it's more than annoyance; maybe it's anger. Now, step back for a moment. We're talking about instant access to an ever-expanding world of information (in a place that was once solely reserved for drinking coffee, having conversations, reading a book or the paper, etc.), using a device with so much computing power that just a few decades ago it would've required tens of thousands of square feet of space. And now that you've got to go back to simply enjoying a cup of coffee and sitting alone with your thoughts for a while, a borderline emotional breakdown ensues.

The digital age has made our lives better in some ways, but it certainly has not made us better. It can't. As Crouch puts it, "Technology is a brilliant expression of human capacity. But anything that offers easy everywhere does nothing (well, almost nothing) to actually form human capacities."[6]

Our growing impatience is a prime example of this. The speed of the digital age is damaging, even destroying, our ability to wait patiently and to live with a long-term perspective. We admire the hare's speed and we mock the tortoise's slowness, forgetting that in the end, steadiness wins the race.

SHALLOW

Because we lack steadiness, patience, and long-term perspective, we're dangerously susceptible to the allure of quick-fix, dopamine-inducing digital experiences. Coupled with the plethora of available choices, we live in constant risk of spiraling into the abyss of our devices. According to professor and author Adam Alter, the average phone usage among adults rose from eighteen minutes per day in 2008 to two hours and forty-eight minutes per day in 2015.[7] This isn't because we're talking to each other more. On the contrary, we're talking to each other less. The dramatic increase is due to emails, internet use, and, in large part, social media. Recent estimates are that there are over three hundred million active users on Snapchat, three hundred thirty-five million on Twitter, one billion on Instagram, and more than two billion on Facebook.[8] (Almost) everyone everywhere is on social media—especially millennials and Generation Z.

All our toggling back and forth on our various social media platforms isn't just making us impatient. It's also making us *shallow*. The fast-paced (the scrolling prowess of our thumbs is at an all-time high), easy-access (our phones are within reach just about anywhere we go), and endlessly customizable (we follow and unfollow who we want, when we want) world of social media is stunting our ability for the sort of depth that Christian discipleship requires.

In his book *Deep Work*, Cal Newport gives this sobering warning: "Spend enough time in a state of frenetic shallowness and you permanently reduce your capacity to perform deep work."[9] The digital age entices and invites us into this never-ending stream of "frenetic shallowness." Scroll, look, like, comment, judge, envy, repeat. It's fast, it's quick, it's easy, and it's often thoughtless and careless. It's shallow and directly counterintuitive to the deep work of discipleship.

Its most insidious effect on us, however, may be the way its constant presence in our lives actually rewires our desires. Based on significant research, Newport concludes that "once you're wired for distraction, you crave it."[10] If we're not careful, social media will change not only our ability but also our appetite. We begin to crave shallow experiences. As the writer James K. A. Smith reminds us, what we crave, what we desire, shapes our identity.[11]

C. S. Lewis's words ring truer than ever: "We are half-hearted creatures, fooling about with drink and sex and ambition when infinite joy is offered us, like an ignorant child who wants to go on making mud pies in a slum because he cannot imagine what is meant by the offer of a holiday at the sea. We are far too easily pleased."[12] The shallowing effect of the digital age isn't just about how we behave; it's about who we're becoming. And who we're becoming in the digital age are shallow shells of the fully alive human beings God designed us to be.

ISOLATED

As the speed and choices of the digital age send us hurling toward impatience and shallowness, they culminate in its most damaging consequence: *isolation*. Social media in particular lures us in under the guise of connection, but beneath this mask is the reality that social media, and digital spaces as a whole, are for the most part lonely places.

This is because social media is fueled by voyeurism—that broken inclination within each of us to peek behind the curtain of other people's lives. Rather than connecting us, the voyeuristic nature of social media actually detaches and distances us from one another, as we find ourselves running aimlessly on the treadmill of comparison and contempt. We feel like we can see one another's lives, but none of us ever feel truly seen. Digital connections often act as poor disguises for our real-life isolation. Sherry Turkle says it this way: "Networked, we are together, but so lessened are our expectations of each other that we can feel utterly alone."[13]

True human connection is fueled by empathy—the God-given ability to step into another's shoes and open ourselves up to another's story, not to compare and contrast, but to be overwhelmed by compassion, to "rejoice with those who rejoice; mourn with those who mourn" (Romans 12:15). This requires patience, depth, and the risk of stepping into real community with real people and their real lives in real time and in real space.

At their best, social media and other digital spaces can be wonderful initiating spaces that lead to true human connection, but they can never become home for those connections; they'll always fall short and leave us wanting. When I FaceTime with my wife and kids (our digital gathering space when I'm away), it's a wonderful benefit of technology—but ultimately it only makes me eager to get home and give them real hugs. That's digital at its best—increasing our appetite for the real, analog thing.

At their worst, social media and digital spaces create a false sense of connection and a façade of community. And they are very skilled at their ruse. (More on these specific ideas in chapters 5 and 6). We must never forget that they are what

Dallas Willard called "dreary substitutes in the form of pleasures."[14] We must never lose our appetite for the real analog thing—true human connection and community, driven by empathy. Without it, discipleship to Jesus just isn't possible.

THE CHURCH IS NOT IMMUNE

In the fifteenth century, when Johannes Gutenberg invented the printing press, the shifting culture began to change the church in surprising ways. As Bibles were placed in the hands of everyday people, the print age reshaped the nature of Christian experience around the book. The life of the mind was more greatly emphasized and biblical scholarship became increasingly important. It is not mere coincidence that some of the most influential early works of systematic theology (e.g. Philipp Melanchthon's *Loci Communes* and John Calvin's *Institutes of the Christian Religion*) were first produced during the print age. While Thomas Aquinas's *Summa Theologia*, another key work of systematics, was written earlier in the thirteenth century, its widespread impact wasn't felt until Gutenberg's press gave it the necessary medium for distribution to the masses.

As the print age shifted the emphasis to the intellectual mind, the sermon began to take on a more centralized place in the worshiping life of the church. This in turn led to the addition of regular seating areas in church sanctuaries, as people were required to sit for longer periods of time. Most often, the seating areas comprised linear rows of pews and an aisle down the middle. Previously, seating had typically been limited to benches around the outside of the room, reserved for the elderly and the sick. It's worth considering if even this physical change that occurred in church buildings—with the pews resembling lines of text on a page and the aisle reflecting a book's spine—were not somehow, at least on a subconscious level, influenced by the emergence of books brought on by the print age. What is clear is that the print age's heightened emphasis on the intellect, which led to the elevation of the sermon as a central element of Christian worship, quite literally changed the way Christians gathered. As *The Oxford History of Christian Worship* notes, "For a thousand years or more they had been on their feet; now their attention was fixed in a single direction. The nave, which had been entirely movement space, now was mostly seating, with movement limited to the aisles."[15] The technological influence of any given age has an undeniable impact on the church.

This becomes even more evident when we fast forward to the mid-to-late twentieth century. Television sets began to dominate American homes and gave rise to the broadcast age. Right around this time, church buildings and sanctuaries began to resemble television studios—big stage, bright lights, multiple screens, and seating arranged for audiences rather than a community of congregants. As thirty- and sixty-minute television broadcasts filled the airwaves, we began to see similar segmentation methods influence the way churches programmed their gatherings together—an opening fifteen-minute segment of songs, a five-minute segment of announcements (akin to commercial breaks in television), and so on.

If the print and broadcast ages each had such sweeping effects on the church, we'd be either foolish or in denial to think that the digital age isn't affecting us in similar ways. This has always been the case. The particular age we find ourselves in always shapes the church. This isn't all bad. The print age and the broadcast age both added tremendous benefits to the church and her effectiveness for mission in the world, both in their time and today. But one of the uniquely dangerous threats of the digital age is that, unlike the print and broadcast ages, it is a non-spatial reality. The internet isn't in any fixed location and it isn't embodied in any set physical medium. The print age was about books, which you could hold in your hands and read aloud for others to hear. The broadcast age was about television sets in the living room, where you'd gather together at set times during the week to watch specific programs.

But the digital age is about speed, choices, and individuality. Fixed locations and physical mediums are seen as impediments to such values. We don't hold it in our hands, we don't read it aloud to one another, we rarely even gather at set times to watch things together anymore. The digital age is about getting what we want, when we want, how we want, and as much (or as little) as we want. And its ill effects are either going unnoticed or are being intentionally ignored by the church, and this is catalyzing a dangerous shift in our ecclesiology.

I was recently listening to a popular church leadership podcast. The host was interviewing a pastor at one of the largest, most influential churches in America. The main point of the discussion was that if churches do not lean into the "digital revolution," they're missing out on one of the greatest gospel opportunities before us and will eventually be left behind in the dust. As I listened, I was struck by the onslaught of parallels between digitally savvy churches and digitally savvy businesses. Multiple times, the pastor mentioned companies like Amazon and Uber. I found myself responding out loud in my car: "Amazon and Uber are products- and services-based organizations. You log on, purchase a book or schedule a ride, and then you log off. You get in, get what you need, and get out. Is that what the church is?"

Almost every argument I've ever heard in support of churches leaning more heavily into the digital age makes this point: the fastest growing, most profitable companies in the business world are the ones leveraging digital platforms. This argument would make all the sense in the world if the church had the same goals as companies like Amazon and Uber.

In the podcast interview, the pastor declared that "digital presence actually drives up in-person engagement." Because we're still in the early stages of the intersection of the church and the digital age, there aren't strong metrics to support or refute this claim on a national level. I assume this is accurate at least in his context and, along those lines, I agree that digital presence is indeed helpful as a front door, so to speak. But at my house, while our front door is important, we place exponentially more emphasis on our kitchen and our living room. Why? Because that's where the truly meaningful connections happen. That's where we create and experience community as a family and with friends who visit us. Our front door is for kindly saying no to solicitors and waving hello to the delivery guy when he drops off a package. But for family and friends, our front door is simply the quick entry point to much more important spaces, like the kitchen table where we share a meal or the living room sofa where we unwind and reconnect.

Leading our churches headlong into digital spaces in hopes of creating an easy-to-consume Christian product severely diminishes our ability to meaningfully impact the culture around us and invite them into more meaningful spaces. The church was never meant to be a *derivative* of the cultural moment but, rather, a *disruption* of it. Amid today's onslaught of digital distractions, the analog church is exactly the sort of disruption we need most to be effective in our cultural moment. As Alan Noble writes, "The greatest witness to the world will always be the body of Christ *gathered* to worship, which means that churches and denominations need to consider well what it means to bear witness in a distracted, secular age"[16] (emphasis mine). The greatest disruption that every church, small or large, can offer is an uncompromising invitation to the kitchen table, the living room sofa, the warmth of the sanctuary, the conversations in the courtyard—the spaces people are truly longing for in the midst of their speedy and impatient, choices-laden and shallow, individualistic and isolated lives.

GOING ANALOG

The Christian church has always been marked by her ability to create and invite people into *transcendent* spaces and experiences. The church has always been most dynamic and effective when she has stood in stark contrast to the dominant culture of the day—zigging when the world is zagging. This sort of creative resistance and prophetic posture is what we need most in the digital age. And the most creative, prophetic way to stand in opposition to the digital age is to lean into analog opportunities.

To gather when the world scatters.

To slow down when the world speeds up.

To commune when the world critiques.

As we serve and lead in the local church, we must remember that the goal isn't selling a product or service but discipling our people. And discipleship requires patience, depth, and community—the very things that stand in contradiction to the values of the digital age. Dallas Willard reminds us that "character is formed through action, and it is transformed through action, including carefully planned and grace-sustained disciplines."[17] Carefully planned and grace-sustained disciplines. This is intentional, methodical, slow and steady work. It's why Jesus used metaphors like vines and branches to describe the life of discipleship:

> I am the vine; you are the branches. If you remain in me and I in you, you will bear much fruit; apart from me you can do nothing. If you do not remain in me, you are like a branch that is thrown away and withers; such branches are picked up, thrown into the fire and burned. If you remain in me and my words remain in you, ask whatever you wish, and it will be done for you. This is

to my Father's glory, that you bear much fruit, showing yourselves to be my disciples. (John 15:5-8)

No matter the fruit, it takes a while for branches to produce it. It requires constant care, regularly scheduled watering and pruning, and daily upkeep. The invitation to remain in Jesus is an invitation into this sort of work, distancing ourselves from the frenetic shallowness of our digital distractions in order to learn and practice the way of Jesus in the big, little, and everything-in-between aspects of life.

I believe there is tremendous opportunity for this, especially with younger generations. Despite the grim news of declining church attendance and engagement among young people, we are also beginning to see these same young people intuitively recognizing and responding to the digital tensions of our day.

David Sax's fascinating book *The Revenge of Analog* presents a variety of ways younger generations are growing more interested in non-digital stuff. Things like Polaroid cameras and moleskin journals are experiencing a renaissance. One of the clearest examples of the analog comeback may be the resurrection of vinyl records. Vinyl record sales have grown from less than one million in 2007 to more than twelve million in 2015, with an annual growth rate of more than 20 percent.[18] In the book, Sax quotes Jay Millar, the former director of marketing at United Record Pressing, as saying, "Digitization is the peak of convenience, but vinyl is the peak of experience."[19]

According to many tech industry experts, the meteoric rise of Amazon signaled the end of bookstores. But this has not been the case. Amazon recently announced plans to open three thousand brick-and-mortar bookstores by the year 2021.[20] Why? Because, as Millar suggests, while buying a book online is *convenient*, it cannot offer the *experience*. Younger generations, having grown up in an over-digitized world, feel this on an intrinsic level and are seeking out experiences they can see, hear, feel, and touch. They realize that ordering a book online and walking through a bookstore are two palpably different things. They're longing for analog. And this offers the church a never-before-seen missional opportunity, to provide these sorts of transcendent spaces that are so few and far between in the digital age.

SEEING THE UNFILTERED SEA

Several years ago, Jenny and I took an anniversary trip to a sleepy little town up the coast of California called Mendocino. We booked ourselves a room at a bed and breakfast that overlooked the Pacific Ocean. When we arrived at the front desk to check in, I asked the customary question, "What's the Wi-Fi password?" The lady gave me a knowing, *oh-you're-one-of-those* smirks. I was confused. She replied, "There's no internet here. Honestly, you're probably not going to have very good cell service either." At first, I thought she was kidding. Then, she pointed to the table in the corner of the room and said, "We do have some great board games right under that table. Feel free to take a few."

I felt the withdrawal symptoms immediately. We were going to be in Mendocino for several days. How would I possibly make it that long without checking email? How would

I survive without knowing what was trending on Twitter? How would anyone know what a great time we were having if I didn't post dramatically filtered photos on Instagram?

Jenny and I got to our room, set down our bags, opened a bottle of wine, and stepped out onto the small balcony. The view was breathtaking. Nothing but the big blue ocean for as far as our eyes could see. We sat in stunned silence for a while, then began to talk. No phones, no laptops, no Wi-Fi, no social media. Just us and the endless sea, seen clearly with our very own unfiltered eyes. Slowly but surely the anxiety of digital disconnection began to fade, and I started to feel an aliveness I hadn't felt in a long while.

This is the opportunity and the challenge before us today as we serve and lead our church communities—to help people lift their collective gaze away from the abyss of their digital devices and spaces, to see Jesus out on the water, inviting them to step out in faith, one small step at a time, to go about the patient and deep work of following him, together.

PART 1

WORSHIP

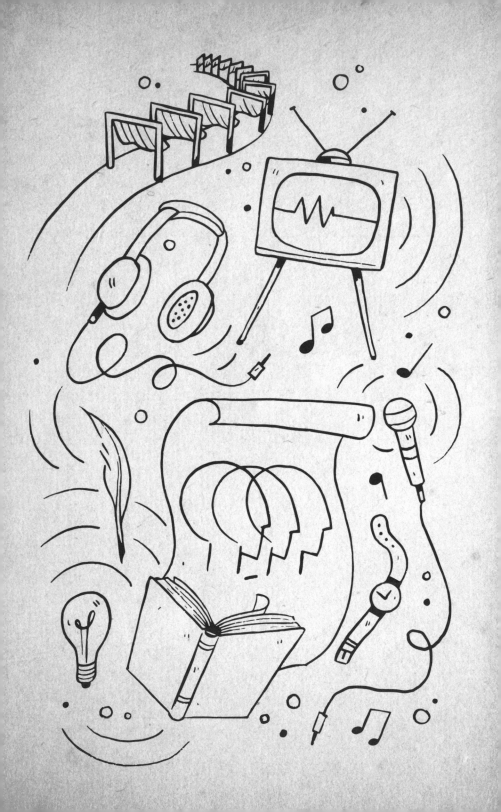

TWO

CAMERAS, COPYCATS, AND CARICATURES

WORSHIP IN THE DIGITAL AGE

Worship is no longer worship when it reflects
the culture around us more than the Christ within us.

A. W. TOZER

A **COUPLE OF YEARS AGO,** I was scrolling through the reviews on our church's Yelp page and came across this comment from a young woman who goes by "Me c.": "If I'm going to roll out of bed on a Sunday morning, I'd rather hear a cool old pipe organ and some good vocal harmonies than someone's extended electric guitar solo."

I felt defensive at first glance, but my defensiveness quickly went away on a careful reading of the rest of her review. First, Me c. doesn't consider herself a Christian, but her boyfriend does, and he wanted to "try going to church on Sundays again." Second, she visited our church more than once ("We went here for a while," as she puts it). Third, she considers herself an "outsider" (her own words). And lastly, Me c. and her boyfriend eventually moved on because they got tired of the "rock concert atmosphere" (once again, sadly, her own words).

On one hand, this is just one person with one opinion. This story doesn't mean that all unchurched people dislike guitar solos. It certainly doesn't mean that there's no place for guitar solos in the church's worship. But this story does offer an important lesson, particularly for those of us who serve and lead in the local church. We must regularly consider and reconsider what we're doing, why we're doing it, and how it's being experienced by the people we're serving and reaching. We must be willing to honestly address the idealized versions of our ideas in our often excessively optimistic minds.

I'm grateful for Me c. and her honesty. This sort of raw feedback can be difficult to hear, but it offers a perspective that I don't have. I've been a part of the local church most of my life, and for the past fifteen years most of my waking hours have been immersed in local church ministry. Without some pushback, I have a very limited understanding of what the local church, and specifically my local church, looks, sounds, and feels like to "outsiders" like Me c. This is an important problem to be solved because "outsiders" like her are people we're deeply interested in reaching and serving.

And what exactly was her experience? She's an unchurched woman, likely in her twenties at the time, based on her profile, with a willingness to at least try. She gave us a chance. She took the risk and spent several of her Sundays with us. For all of our well-intentioned preparation, the environment we created in hopes of reaching and serving people just like her actually pushed her away.

In the next couple of chapters, I want to consider and reconsider some common philosophies and practices when it comes to the way many of our churches shape the substance and style of our worship gatherings. Specifically, I want to consider how the digital age and technology's influence have subverted much of what the worship life of the gathered church is meant to be and how we can move forward in a way that more accurately reflects the worshiping call of the Christian church while also helping her more effectively reach new generations.

WHOLE-BODY PARTICIPATION

If you've been around the church for any length of time, you've heard some version of the idea that *worship isn't singing*. This is undoubtedly true. As Daniel Block writes, "The words that are usually translated as 'worship' in English [Bible] translations have little to do with either praise or music, as today's popular Christian culture suggests."[1] This idea can be extended to include preaching/teaching, as well as less noted yet common elements of typical worship gatherings across America, such as offering and the oft-maligned meet-and-greet time.

But it's also important to remember that worship is not *not* these elements either. While music, preaching, teaching,

Communion, giving, and greeting one another do not en-
compass the whole of biblical worship, they are undeniably
key elements of it. Specifically, they are some of the most
common elements of the worship life of the gathered church.
As we turn our attention to them, we'll consider songs (music
and singing) and sermons (preaching and teaching) in particular (as well as Communion, at the end of the book), because the digital age has impacted and influenced these key
elements in unique ways. In order to get there though, we
have to begin at the beginning—with a basic understanding
of what the Bible says about worship.

In the Old Testament, a few different words are commonly
translated into the one English word *worship*. Two of the
most prominent ones are the Hebrew words *shakhah* and
hishtahavah. *Shakhah* literally means to "bow down" or to
"prostrate oneself." In several places, depending on the specific English version, the word is also translated "fall down,"
"crouch," and even "made to stoop." *Hishtahavah* is a word
used to describe subjects falling prostrate to the ground
before a superior, a posture that would have stated the
bodily equivalent of the phrase "Long live the king!" in the
ancient world. Depending on the English translation, both of
these words are translated interchangeably as "worship" or
several other words or phrases depicting some form of
bowing or kneeling down before a superior.

The New Testament follows suit. The original Greek word
for "worship" is *proskyneō*. The most direct literal translation
of the word means to "kiss the hand," specifically as a sign of
reverence and adoration. This feels antiquated to us in the

modern world, but it was a common gesture in the ancient near east. The word also means to "kneel down" and "touch one's forehead to the ground," both of which were also common gestures of reverence in the ancient world. For example, in Mark 5:6, we read about a demon-possessed man who ran toward Jesus in fear and "fell on his knees" before him. This is the word *proskyneō*. Worship.

In the original Hebrew and Greek texts of the Bible, worship explicitly communicates a *whole-body participation* in reverent response to God. Worship implies bowing down, falling prostrate, kneeling low with heads to the ground, drawing near and kissing the hand, etc.—all acts of adoration and allegiance, all acts that required participation with one's entire body. It's no wonder then that in his letter to the Christians in Rome, Paul instructs us this way: "Therefore, I urge you, brothers and sisters, in view of God's mercy, to offer your *bodies* as a living sacrifice, holy and pleasing to God— this is your true and proper worship" (Romans 12:1).

The original Greek word for "bodies" here is *sōma* but there are a variety of Greek words for "body," and on a sliding scale *sōma* covers the most ground and is the most comprehensive. It means more than simply physical flesh (*sarx*) and it means more than simply unembodied spirit (*pneuma*). The word *sōma* describes both the physical and spiritual essence of our beings. And it's the word most commonly used to describe the church in the New Testament.[2] Does it get any more analog than this? When we read Romans 12:1 within the context of Paul's entire letter, it's clear that he's talking about something entirely more holistic than singing a few songs on Sunday. For

Paul, this call to offer our bodies as living sacrifices is a direct response to the acknowledgment that God is at the center of it all. In the passage that precedes this description of true and proper worship, Paul reminds us of these truths:

> Oh, the depth of the riches of the wisdom and knowledge of God! How unsearchable his judgments, and his paths beyond tracing out! Who has known the mind of the Lord? Or who has been his counselor? Who has ever given to God, that God should repay them? For from him and through him and for him are all things. To him be the glory forever! Amen. (Romans 11:33-36)

Ponder that for a moment.

From him . . .

And through him . . .

And for him . . .

Are all things.

Singing songs is a wonderful expression of worship, but it's clear that songs alone are not enough. Certainly not when we consider that worship is a response to the vast, incomprehensible idea that all things are from God, through God, and ultimately, all things are for God.

All things. Not just music. Not just singing. All things.

Our time.

Our energy.

Our resources.

Our hearts.

Our minds.

Our bodies.

This is worship. So then, is it any surprise that from beginning to end, the Bible's understanding of worship is *whole-body participation*? If all things are from God and through God and for God, the logical, appropriate response would be to give all of ourselves back to him; whole-body participation which acknowledges that none of this is from us, none of this is through us, and none of this is for us. God is the centerpiece, the main event, the bottom line, the beginning and the end, the first and the last. And as such, he deserves everything we have. Not just a few songs. Not just our attention for a few brief moments during the sermon. Everything—heart, mind, body.

What does this mean for the environments and spaces church leaders are called to create to help the people in our communities engage in whole-body participation as true and proper worship? Before we explore what it does mean, we have to first spend some time exploring what it doesn't mean. We have to honestly assess where we may be getting things wrong by unintentionally creating environments and spaces that are counter to the sort of whole-body participation the Bible describes as worship. Specifically, we need to consider the ways our ever-increasingly digital world and its many offerings are reshaping how we worship in our churches today.

McLUHAN'S FOUR LAWS

About half a century ago, there was a prominent philosopher named Marshall McLuhan. The height of his popularity was in the 1960s, but by the '70s he'd lost influence and for decades he was all but forgotten. In recent years, however, parallel to the rise of the internet, McLuhan's work has been making a comeback. This reintroduction is due in large part to a series of shockingly precise (some would say prophetic) claims he made about where he thought technology and media were headed in the future. You can find footage online of McLuhan describing commerce platforms that would arise in the future that sound eerily similar to what we have in Amazon and eBay today. Some say McLuhan predicted the internet itself.[3]

One of McLuhan's most well-known concepts was what he called the Four Laws of Media. By "media" McLuhan meant all extensions of human capacities. He wasn't just talking about CNN or Fox News, as we often think of "media" today; he was talking about anything from the news (which extends the human capacity to access information) to cars (which extend the human capacity to walk and run). His Four Laws of Media can be summarized as a series of questions to be asked about any and every form of media:

1. What does it enhance, improve, or make possible?

2. What does it push aside or make obsolete?

3. What does it retrieve that was previously pushed aside or made obsolete?

4. What does it turn into when pushed to an extreme?

Take the smartphone as an example. The smartphone *enhanced* the human capacity to communicate—specifically, to talk and to listen to one another. It also *made it possible* to access information at any time, with its constant connectivity to the internet. It *pushed aside* mobile phones and made home phones and payphones *obsolete*. Of the many things it *retrieved*, an obvious example is the camera; previously, even those who owned cameras, whether film or digital, rarely carried them around constantly, only bringing them along for specific occasions. But smartphones made photography and video capture a common, everyday occurrence for the masses.

Regarding the last question of the Four Laws, McLuhan stated that when a form of media is pushed to its extreme, it eventually reverses in on itself and works in direct opposition to the very human capacity it was originally intended to enhance. So in the case of the smartphone, its ability to extend the human capacity for communication (to talk and listen, while also being able to see one another, even across great distances) has reversed in on itself. It's now commonplace for people to be sitting right in front of each other, at restaurants and coffee shops and dinner tables, and be totally disconnected from one another, not talking, not listening, not seeing, not communicating, and instead tethered to their devices.

McLuhan's Four Laws of Media have an uncanny ability to shed light and provide perspective on even the most seemingly innocuous technologies. It's a jarring but necessary way to thoughtfully consider what we do and why we do it. Not only do McLuhan's Four Laws reveal the effects of any particular form of media, they also unveil the harm we cause to the very human capacity that that particular medium is meant to help. And in doing so, they can be a catalyst for urgent course correction when necessary. As McLuhan writes, "When the technology of a time is powerfully thrusting in one direction, wisdom may well call for a countervailing thrust."[4]

This is true when it comes to the way the digital age and its technological tools are changing how the church gathers to worship. We need a "countervailing thrust," a shakeup of the way we think about, plan, and engage in worship—especially when it comes to songs and sermons.

"THE 'AMERICAN IDOL' STUFF"

It's believed that the first Protestant hymnal was published in Bohemia (modern-day Czech Republic) during the early sixteenth century by the Unitas Fratrum, later called the Moravian Church. Before hymnals, both the lyrics and melodies of songs had to be memorized, which limited the number of songs that a church could regularly sing together. The introduction of hymnals changed that. Very quickly the church began to see a dramatic increase in the number of songs being written. By the eighteenth century, Protestant hymnody had exploded: Charles Wesley alone wrote more than six thousand five hundred hymns (some estimate he wrote as many as nine thousand).[5] While the hymnal allowed the church to greatly expand its repertoire of songs, it focused the collective gaze of the community downward. In some ways, a sense of togetherness and belonging was lost as eyes became affixed to the words and notes on the pages of individual hymnals held in individual hands. However, this loss was mitigated by the added benefit of being able to sing a wider variety of songs in unison, often skillfully with harmonies, as the melodies and various harmonic notes were written on the pages. Once you knew how to read the musical notes, you could sing along confidently to any song in the hymnal. Many songs eventually became known by heart based on repetition of singing, not only in church but also with family, as hymnals were as commonplace as Bibles in many Christian homes.

In the mid-to-late twentieth century, things changed with the introduction of overhead projectors. These magnificently clunky contraptions allowed for lyrics to be projected onto a wall or screen. Eyes were lifted off of individual hymnals and fixed collectively onto a singular set of lyrics projected at the front of the room. The overhead projector gave us back some sense of togetherness in worship gatherings, as we could see one another, at least peripherally, even while focusing on the lyrics above. But in making the hymnal obsolete, it made it more difficult to sing skillfully and harmoniously, as overhead lyric projections did not include musical notes. The only way to sing well together was to become familiar with the songs over time. We returned in some ways to the time before hymnals, when songs had to be memorized and the divide between the well-rehearsed singers/ musicians on stage and the rest of the people in the seats began to grow.

In recent decades, overhead projectors have given way to their sleeker, sexier cousin, the digital projector. First Power-Point and now programs like ProPresenter have taken lyric projection to new levels. While overhead projectors displayed only the lyrics on clear transparency slides, digital projectors have given us the ability to add graphic backgrounds. When tempered and tasteful, accompanying imagery to song lyrics can be helpful and engaging. They can create a sense of awe in the presence of God, as well as a sense of belonging in the presence of his people. But in many churches, this technology has been taken to extremes and we are seeing exactly what McLuhan's Four Laws of Media predicted—a reversing-in on itself, a movement in direct opposition to the very thing hymnals and lyric projection were intended to do. Namely, to bring us together. As Tim Challies notes, "That little change from book to screen changed nearly everything."[6]

In addition to the pre-existing challenge of singing skill-fully and harmoniously together, the gaudiness of digital lyric projection in many churches today is further widening the gap between those on stage and those in the seats. As background graphics become increasingly more dynamic, and in turn, more distracting, they overwhelm the senses and grab more of our attention than they should. Rather than accentuating the lyrics we're being invited to sing together, these image backgrounds often become a mesmerizing show accentuating a musical performance, and we end up watching rather than participating. We lose our sense of togetherness in the singing life of the church, as the variance between the worship band and the congregation, in skill and familiarity with songs, further accelerated by the spectacle of graphic backgrounds, grows. Rather than a single church community singing together, we are increasingly becoming a gathering of audiences who watch the professionals perform.

Even the way we light our gathering spaces has felt the effects of this unfortunate reversal. With the rise of overhead and digital lyric projection, worship gathering spaces began to darken. Certainly, worship gatherings were not always held in brightly lit spaces before lyric projection, but its introduction made a darker environment not only a matter of choice but of necessity. And as gatherings became darker, supplemental lighting became more prominent. While initially a matter of function, in many churches, lighting quickly began to morph into and mirror the performance venues of, first, the broadcast age, and now, the digital age. Much like lyric projection, lighting can be a powerful tool for connecting people to God and one another when leveraged appropriately. Tastefully done, appropriate, and skilled lighting dynamics can help accentuate a sense of transcendence. But in recent years we've seen church lighting also taken to extremes, with many church gatherings resembling not only television studios but also rock concerts or nightclubs. This unabashed pursuit of bigger and brighter is a chasing after relevance, not transcendence.

We are now forced to consider the unintended impact of the reversal effect of lyrics and lighting taken to their extremes, particularly when we consider how these technological and digital realities in our churches are being perceived by new generations, both churched and unchurched alike. Alan Noble writes, "The lighting and volume make it clear who the congregation should be paying attention to"[7]— namely, the singers and musicians on stage. What were originally intended to connect us more meaningfully to God and one another have now become impediments and obstacles. The shine and sparkle of lyric projection and flashy lighting in our churches are actually repelling the new generations they were meant to reach.

Matthias is in his late twenties. I met him at a U2 concert. Our initial small talk turned to conversation about what I did for a living. When he learned that I work at a church, Matthias began to share honestly about his church experience. He said he'd been at the same church most of his life. He loved it as a kid but shared that in recent years, he'd been turned off by what he called "The 'American Idol' stuff." He went on to explain: "It just seems like everyone's trying to look like they're movie stars up there." I know the church he's from. I know their leaders. They love Jesus and are trying their best to serve their church. They're not trying to be movie stars. They're just doing what they think they're supposed to do, because all the big churches seem to be doing it too.

Patrick is a young man in his twenties. He's from the Midwest but spent a summer in Santa Cruz, and while he was here he attended our church. He isn't a Christian, but he finds the life and teachings of Jesus intriguing. After a few months with us, he was set to move back home and asked me if I could recommend a few churches in his hometown that he could check out. I've never been to the place he's from so all I could offer were a few churches I found online that seemed like healthy church communities. After a few weeks, Patrick emailed me to share about his experience at the first church I suggested, which employed the sort of lyrics-and-lights approach I'm writing about here. He said, "It felt commercialized and kind of impersonal. The service was intense and felt like a rock show." Then he added, "And the sermon was projected on a screen, like a video of an offsite pastor. It was strange. What's that about?"

Great question, Patrick. What *is* that about?

TALK TO THE CAMERA

In the introduction, I briefly mentioned a moment when I was on staff as one of the teaching pastors at a multisite church in the Silicon Valley. As I was readying to step up on stage and deliver a sermon, the service coordinator reminded me, "Jay, don't forget to look directly into the camera at the back of the room so the campuses feel connected to you." The thought of looking into a camera to "connect" with people who would be gathering on another day in another room on the other side of the city struck me as an exercise in missing the point.

In recent years the multisite church model has become the go-to strategy for church growth. Many large churches are moving in this direction, where the key elements of the church are replicated and repeated in new sites, akin to the franchise model of places like Target. When you walk into a Target store in California, its look and layout is extremely similar if not completely identical to a Target in Virginia. And so it goes with most multisite churches. While many have experienced some growth over the years, the model is beginning to reveal cracks in its armor, as younger generations find the franchise approach lacking the sort of uniquely contextualized and personalized approach they see as integral to authenticity and creativity. One of the major factors fueling this dissatisfaction is the experience of sermons on screens.

As part of their commitment to maintaining brand consistency, many multisite churches employ a video-teaching model. While the other elements of the worship gathering are usually live and in person, most multisites allot thirty to forty minutes of their gathering time to a video of the teaching, sometimes streamed live and sometimes prerecorded, from whichever site the communicator happens to be in-person that weekend (typically, the largest site). By their very nature, video teachings cannot effectively engage the nuanced stories and cultures of the people they seek to serve. As professor and author David Fitch writes,

> [Video teaching disregards the] local context, its culture and instead assumes that who we are and what we say as a church applies to you with no dialogue or presence needed. It asks people to come to me on our terms. . . . The result is that this church most often will 'attract' people of like ilk who already believe the same things and use the same language to gather in a homogenous group. . . . When this happens, this church has become incapable of mission.[8]

How did we get here? Once again, applying McLuhan's Four Laws of Media is helpful.

When pulpits were first introduced into worship gatherings, they served a two-fold purpose. The first was theological. The Word of God was about to be preached. This was and is a sacred act, a moment in which the collective people of God open their hearts and minds to receive the transformative grace and truth of the unfolding story of God together.

The second was pragmatic. While pulpits would eventually evolve to take on a variety of shapes and sizes, the earliest pulpits shared the commonality of being raised and enclosed platforms from which the preacher would deliver the sermon. This allowed people to see the preacher more clearly, with the hope being that they'd connect more directly and listen more intently because of the visibility.

Similar to lyric projection, video projection of the preacher onto large screens in the sanctuary extended the original intent and purpose of pulpits and raised platforms. Projecting the preacher on the screen made the preacher more easily seen, thus extending the intent and purpose of connecting the preacher to the people on a more direct and personal level.

But as with lyrics and lights, projecting the teacher on video has been taken to the extreme. What was originally intended to close the gap and narrow the distance between the communicator and the community has now worked to create a disconnect wider and broader than ever before. Rather than creating a sense of personal connection and intimacy with the preacher, in many multisite churches people now sit and watch a communicator who can't see them, hear them, or feel their responsive presence to what's unfolding in real time. If we believe that sermons are purely monologue, then maybe this isn't a problem. But I think most of us would agree that at a deep level, sermons are more than monologue. While the listening community doesn't typically engage in active ongoing conversation during the sermon (unless you're a part of one of those really amazing Pentecostal communities with tons of "Amens!" and "Hallelujahs!"

throughout), there is a sort of dialogue happening. As Thomas Long writes in his seminal book on preaching:

> Preaching does not occur in thin air but always happens on a specific occasion and with particular people in a given cultural setting. These circumstances necessarily affect both the content and style of preaching, but if we think of preaching as announcing some rarefied biblical message untouched by the situation at hand, we risk preaching in ways that simply cannot be heard.[9]

Preaching does not occur in thin air. It requires a specific occasion, a particular people, and a specific cultural setting. Then, Long's most sobering reminder: If our preaching is untouched by the situation at hand, we risk preaching in ways that simply *cannot be heard*. Author Jake Meador writes that "sermons assume a personal proximity and a pastoral relationship—or at least the potential for one— between the preacher and the hearer."[10] If we are not in the room, standing in the very midst of the people to whom the sermon is being delivered, if we cannot see their faces, hear their singing, feel their palpable anticipation, need, or yearning, then how can we possibly preach in ways that can be heard?

This is not a blanket critique of the usage of video in churches, as I do think video technology can be leveraged in helpful, supplemental ways. But when tools go unchecked and are used for things they were never intended for, they can cause great damage. A hammer in the hands of a skilled carpenter can be used to build an endless array of wonderful things. But a hammer in the hands of an unskilled, unchecked individual can do great harm.

I love local churches of all shapes and sizes. I want to see them thrive and impact their corners of the world in the unique ways God's called them to. But the digital age and its impact on the worshiping life of the local church has gone unchecked in recent years. This has done great harm to our church communities. It has prevented us from thinking creatively about reaching new generations, who are unimpressed and even repelled by the church's pursuit of digital relevance.

And in addition to the harm it's done to our churches, the unchecked effects of the digital age on the worshiping life of the church are doing damage to the very men and women charged with serving and leading the church into the future. They are doing damage to you—tapping into your insecurities, uncertainties, and performance-driven tendencies in the worst possible ways.

COMPARISON, COPYCATS, CARICATURES

Today people have access to the best sermons and church music in the world, at the touch of their fingertips. So why don't they just stay home and listen from the comfort of their beds on Sunday mornings? For those of us who are leading church communities, putting forth our best effort week in and week out to craft sermons and songs that might impact our communities, how do we deal with the reality that our people can more conveniently stay home and listen to the Tim Kellers, Andy Stanleys, Francis Chans, and Erwin McManuses of the world—gifted communicators whose content is, more often than not, far more engaging than anything we could come up with on our own?

Put your hand out about six inches in front of you, in your line of sight, with your palm facing you and your fingers spread wide. Then, fix your eyes past your hand, to something several feet away. With your eyes fixed past your open hand, describe to me, in detail, the various lines and contours of your palm and fingers. Really, go ahead.

Pretty difficult to do, right?

This is how so many of us in church leadership today are crafting, creating, and leading our worship gatherings. The communities we've been called to serve are right there in front of us, with all of their unique quirks, untapped potential, and original stories. They're brimming with possibility, waiting for creative, visionary leaders to see what most cannot yet see and paint a picture of what could be and come alongside them to make those things a reality. But so many of us aren't doing the work of creatively leading our own communities, because our eyes are fixated on something far away.

We're fixated on what this church or that church is doing and figuring out how we can do that too.

We're fixated on the next hit Christian worship album and figuring out how many of those songs we can play in our church, as soon as possible, so that we too can sound just like Hillsong or Bethel or whatever else.

We're fixated on consuming as many sermon podcasts as we can so that we can teach and preach just like Tim Keller, Andy Stanley, Francis Chan, Erwin McManus, or whoever else.

But these fixations are most often getting in the way. They are sending us spiraling down the path of comparison, which initially turns us into copycats and then eventually into caricatures. And caricatures, by their very nature, have no originality, no creativity, and no faithfulness to the localized culture and context, whatsoever.

I mentioned the multisite church where I served on staff earlier. It's a wonderful church that continues to have significant kingdom impact in its city. Because of the strong reputation of its leadership, built over many years of faithfulness (which began long before I ever got there), it was not uncommon for us to be visited by other pastors in the city. I would regularly meet with other teaching pastors who wanted to talk about our preaching culture and sermon series we were contemplating and to look at some of our content. We considered this a tremendous honor and we were humbled that the things we were working on might help other churches.

But in the midst of the many meetings I had with other pastors in town, time and time again, I would find myself concluding our conversations with these same words, almost verbatim each time: "This might not work for your church. Take whatever's helpful. Toss out the rest." What I've realized in recent years is that I need to hear these words for myself too.

This might not work for my church.

Take whatever's helpful. Toss out the rest.

These are important reminders that help keep us from becoming entangled in the rut of comparison. I'll readily admit that if I'm not careful, I'm quite susceptible to comparison myself. I've wrestled with it my entire life. It's a constant struggle to reorient myself around my identity in Jesus Christ and not around what my life and ministry look like against the glossy backdrops of men and women who seem more successful and more effective than me. The digital age has compounded the problem. Because I have infinite access now to the best and brightest churches and their incredible leaders, I often find myself fighting the incessant temptation to see what they're doing and compare myself and our church to them.

On top of that, I sometimes find myself afraid that the people I serve might be doing the same. They have access to the same content, the same great teaching and preaching, the same great music, and the same great initiatives and ideas from churches all over the world.

My guess is that I'm not alone. My guess is that many of us share these same struggles. But we must remember that these struggles and fears are unfounded. You are not Tim Keller. And your church is not the one you read about in that article. Neither am I, and neither is my church. You're you and I'm me. Your church is the one you were called to serve and mine is the one I was called to serve. It's sort of silly that I'm taking the time to write these words out because it's so obvious, but what's even more ridiculous is how often you and I forget these things. However, this isn't at all surprising because the digital age is designed in some ways to make us forget.

We forget because we're too busy to remember.

We're too busy moving on to the next idea.

We're too busy trying to keep up with the church and ministry Joneses.

We're too busy seeking the next big thing.

And we end up missing the next right thing.

It doesn't have to be this way. Everything we need to fully maximize and realize the potential of our churches is already right there in front us. The key is to fix our eyes away from the wrong things and onto the right things. It isn't about bigger or brighter or louder. It's about discovering the various lines and contours of our people and our places and unearthing the hidden things that hold unknown potential for impact. It's about breaking away from the poisonous spell of comparison, which slowly erodes us into copycats and caricatures. It's about healing from the digital disease which has infected so many of us. It's about moving our gaze away from the distant digital abyss, with all of its shiny, glossy images of what other churches and leaders are doing, and instead, fixing our gaze right in front of us, on the lives of the people we're called to serve, and curating spaces and places of worship that accentuate their unique stories and project a unique vision for who we could become, together.

This is what I call analog worship, and it's what the next chapter is all about.

THREE

TO ENGAGE AND TO WITNESS

ANALOG WORSHIP

Worship reminds us of the shape of true life.

ANDY CROUCH

SPENT A WEEK IN HAITI IN 2014 visiting missionaries who work in some of the most rural parts of the country. A couple of friends and I were there on a fact-finding trip to see if there were specific ways our church could partner with the work they were doing. On our first full day in the country, we woke up at 4:30 a.m. and shuffled into a small van. We drove about half an hour on unpaved dirt roads in the pitch black of night, until we pulled up to one of the smallest villages I'd ever seen.

As we stumbled out of the van, jet-lagged and groggy, we heard what sounded like a thousand voices filling up the darkness. We followed the sound until we arrived at a small dirt patch where about seventy-five people from this tiny little village were huddled together. Mothers and fathers, sons and daughters, grandparents and grandchildren, all gathered to sing before the sun brought its heat. For nearly an hour, they sang. No band, no lyrics projected on a screen, just the sound of many voices, young and old, as one. Song after song, they sang of their desperation and longing for Jesus. The singing stopped only for a few brief moments as one of the men opened his Bible and delivered a short sermon in their native Haitian Creole. The community was engaged and responsive, clapping and cheering to spur the man on. After the sermon, they continued singing. They sang one song longer than the others. For several minutes, they sang the same beautifully melodic words on repeat. The sustained mantra lost no momentum or energy over that time and instead, sounded truer with each passing moment. I asked what they were singing. Our host told us they were singing,

We need Jesus

We need Jesus right now

As they sang these simple and wonderful words over and over, the sun began to rise. I couldn't help but think of Malachi 4:2, "But for you who revere my name, the sun of righteousness will rise with healing in its rays." I sensed with utmost clarity that we were in the middle of something transcendent.

What made this experience so special? How was it that a setting so utterly devoid of the sophistication we're used to in the United States—bright lights and big sounds, technological

advancements and dynamic personalities—could be so undeniably powerful? This wasn't the first time I'd experienced
something like this. I'd had similar moments in other unfamiliar places, from Asia to Africa to Central America. And
while each experience was unique, they all shared a common
equation: minimal resources plus a strong sense of community plus awareness of their need for Jesus equals a transcendent worship experience. After some time distanced me
from each of these experiences, I always came around to the
same two questions: Did the experience simply *feel* powerful
because of its novelty? Maybe the experience would lose its
power with familiarity. Maybe it wouldn't be as good the
second time, like a movie with a twist ending. How were these
mostly untrained and unremarkable musicians and communicators so consistently and effectively captivating the hearts
and minds of their communities?

Both questions can be answered by considering the communities themselves. In Haiti, I was surrounded by men,
women, and children who call that village home. For most of
them, that village will be home for their entire lives. We were
told that they gather like that each morning four or five days
a week to sing, pray, and open the Bible together. Nothing
about the experience is novel for them. This is a part of their
ordinary rhythm of daily life. Yet their passion was palpable.
That transcendent worship experience in Haiti cannot be ex-
plained away by novelty. So what does explain it? This leads
to the answer to the second question. These untrained, unremarkable musicians and communicators are able to cap-
tivate the hearts and minds of their people because they are

their people. Despite the lack of resources, they have something we are sorely missing in the American church; they have a sense of rapport and trust, built on a daily sharing of life within the community. They're *with* one another, in almost every sense of the word, united for the long haul. My brief experience with them reminded me of these words from the German theologian Dietrich Bonhoeffer: "Our song on earth is speech. It is the sung Word. Why do Christians sing when they are together? The reason is, quite simply, that in singing together it is possible for them to speak and pray the same Word at the same time—in other words, for the sake of uniting in the Word."[1]

These leaders are faithful to Jesus, and deeply connected to the people they're serving and leading. This was and is the answer—in Haiti and right here in digital-age America.

DIGITAL INFORMS.
ANALOG TRANSFORMS.

The modern evangelical church in America has been heavily influenced in recent decades by business-leadership models. While there have been some benefits to the organizational infrastructures and practices of our churches, this influence also pushed us heavily in the direction of pragmatism and results-based decision making. A pragmatic approach and goal setting to achieve results are vitally important and they should not lose their place in church leadership. However, we must become more aware of the ways in which they've blurred our vision when it comes to thinking critically about our methodologies.

This pragmatic, results-based approach has led many to a methodology that blindly leverages the latest and greatest technological advancements. But such a digital approach to the worship of the local church more often than not works against the deeper, truer, more meaningful reasons why the people of God have always gathered together to sing, to listen, and to create together. We are in desperate need of recapturing what the men, women, and children in that tiny Haitian village understand—it is trust built on a shared life in the community, a sense of belonging, and shared experiences which usher us into transcendent worship experiences that transform us into the people of God together. This is what I call an *analog* approach to worship.

What I am beginning to see more and more clearly as I consider what is (and is not) happening in the worshiping life of our churches, is this key dichotomy:

Digital informs.

Analog transforms.

The truth is we need both. We need information. But information should always move us toward transformation. Information is the *means*; transformation is the *end*. In the previous chapter, I focused on two crucial elements of the worshiping life of the church—songs and sermons—and how the digital age has taken them to unhelpful and even harmful extremes. In this chapter, we will explore the dichotomy between digital and analog as it relates to these two elements. In order to reclaim the transformative power of songs and sermons in the life of the local church, we must begin by asking two key questions:

When it comes to the singing life of our churches, we must ask the question, "Does this entertain or engage?"

When it comes to the preaching life of our churches, we must ask the question, "Are we inviting people to watch or to witness?"

ENTERTAINED OR ENGAGED

These days, music is everywhere. It's on television and film, elevators and restaurants, public bathrooms and dentist offices. It's in our cars and on our phones. With just a few taps to our screens, we can access almost any song from anytime and anywhere. But it hasn't always been this way. For most of human history, music was a strictly ephemeral experience. Sound was first recorded in the mid-nineteenth century, and it wasn't until the very late-nineteenth century that recorded music as a form of entertainment became normative. Before then, if someone wanted to hear music, they had to find the people who played music and the places where they were playing it. Songs were localized. Songs were events. A song wasn't just about five minutes of sound. A song was about the journey to a location, the gathering of a community, the anticipation of the first note, and the mix of satisfaction and sadness when the final note ended. Songs were moments to be cherished and memories to be captured. Even more than that, songs were an invitation to create these moments and memories together. And for most of history, there was one primary place where this invitation was standing and open, week after week: the church.

For many years the church was the predominant place where entire local communities gathered regularly to create music and sing songs. Those who wrote songs for the church always wrote with participatory audiences in mind. They wrote songs designed not simply to *appeal* to the masses but primarily to *involve* them. In the digital age, we've lost sight of this great heritage. The singing life of the church was built on the foundational principle of engagement—gathering to sing, create, and worship together. But in the digital age, this principle has devolved into an almost unrecognizable (to most of church history, at least) form of entertainment. Andy Crouch, who is not only a writer but also a trained musician and worship leader, describes what we experience in most of our churches today this way: "Our worship bands are more technically proficient than ever, and louder than ever. The people holding microphones are singing, often expertly and almost always passionately. It's just the rest of us who, like the crowd at a ballgame, are mostly swaying along, maybe echoing a few of the phrases or words."[2]

The point here is not that worship leaders and musicians shouldn't be skilled and good at what they do. Bad musicianship and an untrained, ill-equipped worship band can be distracting. The point here is that in order to invite people into analog worship in the digital age, worship leaders and musicians must leverage their skill for something quite different than what we see in the rest of the world of music, where the lights shine brightest on the people in front of the microphones and behind the instruments. We must critically examine and consider if in our churches we've succumbed to the cultural temptation of emphasizing spectacle over substance. In his book *Honest Worship*, worship leader and musician Manuel Luz explains it this way:

> In the midst of all the smoke machines, high-def video loops, and latest worship hits, we may be settling for something less than true transcendence, something less than Spirit-breathed worship, something less than God on God's terms. Are we inadvertently teaching ourselves to settle for spectacle, to be satisfied by titillation—and maybe even become dependent on it—in our worship?[3]

In the singing life of the church, worship leaders and musicians must give greater thought to how every part of our musical worship—song choices, volume, lights, even stage layout and posture—draws people into substance, not spectacle. Everything we do must invite people to engage and participate and not let them off the hook, to simply sit back and be entertained. This sort of analog reorientation of musical worship gets down into the nuts-and-bolts details.

A church just north of where I live has put a lot of intentional thought into this participatory approach that invites engagement. They belong to a network of churches, with local congregations from Honolulu and Los Angeles to London. Even though these churches are all over the world, if you walk into one of their churches anywhere, you'll immediately experience the same backlit lighting on stage during the singing. Because these lights shine on the worship band from behind them rather than in front of them, it creates a very particular mood and, more importantly, communicates a very particular philosophy. The lighting, which silhouettes the people on stage and creates a healthy sense of anonymity, communicates that this experience isn't about the band up front but about us collectively encountering and responding to God together.

This same approach applies to volume and song choices. My friend Jack is in his eighties. He needs a walker to get around, and it's an ordeal, but Sunday after Sunday he arrives at our 9:00 a.m. gathering and sits in the same seat with his wife. He has a hearing aid, which comes in handy when the volume of the music in our gatherings overwhelms him. But recently, after some skillful adjustments made by our sound technician, Jack told me how thrilled he was that he could turn up his hearing aid and hear not only the band but the entire room filled by a sea of voices. He was delighted to be able to sing the familiar hymns as well as the theologically rich and easy-to-sing contemporary songs that morning.

We're beginning to see this turn toward analog worship surfacing in surprising places. New Life Church in Colorado, an evangelical, multisite megachurch of more than ten thousand people, concludes their Sunday gatherings at their large downtown campus by singing the Doxology in acapella every week.[4] Other large, influential churches like Willow Creek in Illinois, Mars Hill in Michigan, and the Village Church in Texas are incorporating more participatory liturgy into the regular rhythms of their weekend gatherings.[5] These communities, sometimes categorized from the outside as "seeker-sensitive" or "attractional" churches, are recognizing the need for a more participatory and engaging worship environment and are making necessary changes.

These are simple yet vital adjustments we can make in our churches to create a more analog worship experience that moves our people from *entertainment* to *engagement*. Moving in this direction begins with recognizing our current deficit and what it is costing us. Andy Crouch writes, "The reorientation of our musical lives around consumption is robbing us of something deeper; it is robbing us of a fundamental form of worship."[6] What he means is this: the "easy-everywhere"[7] access to music in the digital age has led us to believe that music is primarily something to be *consumed* rather than *created*. But it is in *creating* music together that we experience its power most fully. Scientific research is revealing that this is wired into our very DNA. Singing and creating music together has a strong positive effect on physical and emotional health, as well as an accelerating impact on relational connections.[8] God made us to sing together.

Dietrich Bonhoeffer once wrote,

> It is the voice of the church that is heard in singing together. It is not I who sing but the church. However, as a member of the church, I may share in its song. Thus all true singing together must serve to widen our spiritual horizon. It must enable us to recognize our small community as a member of the great Christian church on earth and must help us willingly and joyfully to take our place in the song of the church with our singing, be it feeble or good.[9]

In the digital age, it's easy to lose sight of our place in the larger story and to live with a narrow, self-centered perspective. But when we gather to worship through music and singing, we have the opportunity to widen our spiritual horizon, to willingly and joyfully take our place in the song of the church. This work can only be done if and when we embrace an analog approach to worship, inviting people to engage, to create, to lend us their voices and not just their ears.

WATCHING OR WITNESSING

Several of the local news affiliates in town brand themselves as "Eyewitness News." I'm sure this is true where you live as well. And true to their branding, when you watch one of their 6:00 p.m. broadcasts, there are multiple segments with reporters on location at places of interest. I've wondered what it is about having these reporters on location that matters so much. It costs a lot of money to have a reporter and a camera crew driving all around town, chasing down stories in real places in real time. It'd be much more cost efficient to simply hear about a news story happening someplace and relay the information from the confines of the television studio.

But it *does* matter. The reason we care about the news being delivered by eyewitness reporters is because we want to know that the storyteller is not only *informed* but also *experienced*. It matters to us that the reporter not only knows the information but also that she's experiencing the story firsthand as it unfolds.

As I wrote in the previous chapter, an increasing number of large churches are moving to multisite models of church by utilizing a video-teaching approach. This approach makes all the pragmatic sense in the world, on a number of levels. Everyone at every venue hears the same sermon from the

same person. Staffing hours are not spent on multiple communicators crafting the same basic teaching points. A sem-
blance of real connection can be established between the
entire church community and the primary communicator.

But this approach to preaching fails to address the need
for *witness*. When a sermon is delivered via video, no matter
how dynamic and gifted the communicator may be, the
sermon is inherently a *watching* experience, not a *witnessing*
one. And when it comes to preaching, the difference between
watching and *witnessing* is everything.

Most people today think of sermons as extended religious
monologues meant to encourage, inspire, and occasionally
challenge us. But a sermon is much more than that. A sermon
is a transcendent act intended to transform us, and transformation demands participation—not simply a detached, isolated response after the fact, but actual participation in the
moment. When I go to the gym and meet with a personal
trainer to transform my increasingly unhealthy middle-aged
body, I do not sit on the sidelines and watch him do the exercises and then go home and attempt to replicate the exercises
after the fact. I do the exercises along with him in the moment,
together, which then carries over to continued participation
even when we're apart.

In much the same way, preaching is a participatory act
involving both the communicator and the community, in the
moment, not simply after the fact. Because we've lost the
nuance of participation while listening, sometimes understood as active listening (yet another negative effect of the
digital age), we've resigned sermons to purely informational,

sometimes inspirational categories. But as Thomas Long reminds us, "The sermon is action; the sermon is what the
preacher speaks joined with what the rest of the congregation hears."[10] In other words, the sermon is much more
than the prepared content of the communicator and its
public delivery; it is the sum total of its various elements—
speaking, listening, delivering, receiving, responding—and
it involves everyone in the room.

This is why preaching is an act that must be *witnessed*
rather than simply *watched*. Participation in the transformation process begins at the moment of sermon delivery.
The difference between a local news story and the gospel is
that while news stories have a mostly peripheral effect on
our lives, the gospel is intended to affect us directly, to the
very core of who we are and are becoming. In other words,
while it works well for us to *watch* the eye*witness* telling of
a news story, when it comes to the gospel, we are invited to
be the *witnesses* ourselves. So any time the gospel is preached,
any time a sermon is delivered, the people of God are meant
to gather and *witness* it for themselves. This goes beyond
simple pragmatics and gets into a nuanced yet essential understanding of human experience. There's something
uniquely powerful about the live, in-person experience.

Several years ago, my friend Curtis began posting YouTube
videos of some magic tricks he'd learned. They were unbe-
lievable. So what did I ask him to do the next time I saw him
in person? You guessed it. I asked him to do some magic tricks.

Why? I'd already seen the tricks online. But when something captivates our imagination, we want a better view, a

closer look, a more intimate experience. We don't simply want to *watch* it, we want to *witness* it for ourselves. This is why we pay to attend sporting events that we could watch for free at home or concerts we could watch for free online. In his book on preaching, Tim Keller says that the way to deliver not just an informative lecture but a life-changing sermon is "not merely to talk about Christ but to show him, to 'demonstrate' [1 Cor. 2:4] his greatness and to reveal him as worthy of praise and adoration."[11] It is difficult to *show* and *demonstrate* over a video. It's even more difficult to immersively *experience* and *participate* digitally. Some things—the most transformative things—demand our whole-bodied presence.

In February of 2005, artists Christo and Jeanne-Claude presented their piece called *The Gates* at Central Park in New York City.[12] The couple had been conceptualizing, preparing, and creating the piece for twenty-five years. When it finally opened to the public, twenty-three miles of footpaths throughout Central Park were adorned with 7,503 sixteen-foot-tall gates, each featuring a sheer orange fabric, dancing in the wind. In his review for the *New York Times*, Michael Kimmelman wrote that *The Gates*, "need to be—they are conceived to be—experienced on the ground, at eye level" and that they "beckon people to discover what is beyond them."[13] I've seen many stunning photographs of *The Gates* and I am acutely aware when only looking at them that they fall dramatically short.

The same is true of sermons. They need to be experienced "on the ground, at eye level." They point everyone present to "discover what is beyond them." The potential for human connection through the acts of speaking and listening, delivering and receiving, can be fully realized only when we understand and engage preaching as an act that, in the words of Leonard Sweet, "takes place without audiences, only with participants and partners."[14]

This is not to say that videos should be totally removed from the life of the church. Far from it, actually. Digital technology has given us great advantages over past generations when it comes to the exchange of information. Even in our worship gatherings, videos can be leveraged to accurately and effectively share necessary information about the life of the church. For those of us leading multisite churches, wondering how an analog approach might radically disrupt the way we currently do things, it's important to remember that digital can indeed *inform*. In many ways, digital mediums inform more dynamically than analog ones, as the message can be edited and crafted to be clear and compelling in ways that in-person analog approaches may not. But as I also stated earlier, only analog approaches can truly *transform*. So the critical question to ask when deciding whether to leverage video platforms in communicating a message is this:

Is the goal of this message to inform or transform?

If the goal is to inform, video away! But if it is to transform, then we must lean into analog approaches as much as possible. This will mean equipping more and more gifted men and women to effectively communicate the gospel through preaching. This may also mean reconsidering our staffing and budgeting philosophies. This could even mean prayerfully and thoughtfully assessing if the entire organizational model of our churches may need to shift. I am grateful that several influential churches are already going about this difficult but necessary work.

The Village Church in Dallas, Texas, which I mentioned earlier, is one of the largest churches in the country. For many years they've been led by Matt Chandler, who is widely acknowledged as one of the most gifted communicators in the church today. The Village Church has grown over the years to a multisite church with five campuses throughout the Dallas-Fort Worth area. For many years the church utilized a video-teaching model. In more recent years, it adopted a hybrid model which had a balance of video teaching and live teaching. But in September 2017, Chandler and the elders of the Village Church announced a plan to launch the five campuses into five autonomous local churches, with the goal being to continue church planting efforts, both locally and globally, through these five unique communities.

In explaining some of the story, Chandler describes seeing the fruitfulness of the first campus they launched in Denton this way: "Watching them (Denton Church) really in a contextual place, be freed up to engage where they were, was really a beautiful thing and shone some light on some weaknesses that this thing that we're doing (multisite), it's got an expiration date."[15] Chandler is talking only about the Village Church and not making a blanket statement about multisite churches in general. But what he and the elders of the Village Church came to realize is not unique only to them. There is a growing sense that this may be the way to move forward.

Shortly before the Village Church's announcement, another influential pastor and his church made a similar announcement. Tim Keller planted Redeemer Presbyterian Church in New York City in 1989 and had served as lead pastor for twenty-eight years, watching as the church community grew to more than five thousand people attending weekly Sunday gatherings. But in July of 2017, he stepped down to focus on teaching, writing, and helping to lead a global church-planting movement. In his statement he said this:

> The mission of Redeemer has always been to be a church that serves the city. Starting July 1, we will embody that mission in a new form by actualizing our long-stated plan of shifting from being a single large church with multiple congregations to becoming a family of smaller churches. . . . As smaller churches, they will be better positioned to serve the neighborhoods they are a part of, raise up and train new leaders to serve the city, and send out members to plant additional sites and churches in more neighborhoods.[16]

One of the most encouraging things about the shift we're seeing from large, influential churches and their leaders is its potential for deconstructing the idolatrous influence of Christian celebrity. While the digital age did not create the cult of Christian celebrity, it has certainly enhanced and advanced it exponentially. I am alarmed at the rising tide of self-promotion, salesmanship, and self-aggrandizement that runs rampant among some well-known church leaders. I've never been in the shoes[17] of those who have national and global platforms, so I have no idea about the pressures and temptations that lie in those places. I do know myself well enough to say that if I were in their shoes, the chances are good that pride would often get the best of me too. But in moving to a more analog approach in music, singing, and preaching, we can build a framework to work against the pervading temptation to get beyond our local contexts in order to achieve some sort of national and global notoriety. Using our influence to equip, train, and provide opportunities for other leaders to exercise their own gifts of leadership, whether through preaching or some other means, is one of the most surefire ways to deconstruct the sinful desire within each of us to make ourselves the heroes of the story.

At its finest, the sermon blurs the line between communicator and community, and we experience its transformative power together. As Thomas Long writes, "People may call it 'our sermon,' but it does not belong to us alone. It belongs as well to those who help create it by their listening. To put it theologically, a sermon is a work of the church and not merely a work of the preacher."[18] We must begin to see preaching as an opportunity to embed ourselves in real time and in real space with real people, people we can see, hear, and touch. We must begin to realize that the sermon has little to do with what we can produce or achieve, but rather, that it is a gift of God we are given to steward and to share, both in its crafting and in its delivery. When we can carry the gospel through our words and our posture in the presence of the communities we serve, we jump into the transformation process together.

JOY AND MOURNING, CREATIVITY AND ARTISTRY

As we engage an analog approach to the worshiping life of our churches, there are two specific philosophical shifts that can become helpful lenses through which we consider a wide variety of things we do when we gather to worship.

The first is a shift away from *hype and happiness* and toward an emphasis on *joy and mourning*. In most digitally saturated worship environments today, there is an over-abundance of hype, which often leads to what feels to many like a forced manufacturing of happiness. The University of Washington released a fascinating paper recently called *God is like a Drug*. Based on extensive research done on the gathering methodologies of large, technologically savvy churches, the researchers found that the emotional energy created by these environments lead to the release of extremely high levels of oxytocin (the "happy" chemical). But they warn that these high levels of happiness are "driven on a non-reflective level, and only later does the person 'rationalize' on a reflective, cognitive level, their behavior."[19] In other words, the hype and happiness lead us to feeling great for a little while but not quite knowing why.

The biblical call to worship is much more thoughtful and holistic than that. We are certainly called to engage with God with our hearts, emotions, and feelings, but we are also called to engage with our minds, bodies, and souls, that deep part of us where *all* our thoughts and feelings find their foundation. An analog approach to worship requires us to create space and invite our communities to engage with God fully with whatever is most pressing and pertinent to their lives in the here and now. For some this will lead to great joy. For others it will lead to mourning. For many it will lead to something in between. But wherever it leads, analog worship must create room for it all. The rhythms of our worship gatherings must begin to welcome the wide spectrum of emotions born out of the countlessly unique stories and situations people bring into the room. We must recognize and give voice to joy, mourning, and everything in between when we gather to worship together, acknowledging that God meets us in all of it.

The second shift is a shift away from *digital sophistication* and toward an emphasis on *creativity and artistry*. My friend Karole is a visual artist. She often does live painting in our worship gatherings. I asked her once about why painting live during the songs and sermons at church is a meaningful experience for her. She said this: "Art can speak to me in a language that words cannot reach. I like to see live art in churches because it calls us to be active participants and asks us to witness the creation process. If you are watching me paint, you, along with me, are seeing what God is doing." At our church, we invite Karole and a host of other artists to share their art live in our gatherings regularly. As she says it, "witnessing the creation process" is a much more participatory act than looking at a carefully crafted and edited finished product.

Now, I do think digital media can be a powerful tool. Early in my college years, I strongly considered majoring in film, with dreams of becoming the next Terrence Malick. I think film and video can be used beautifully and in very personal, intimate ways, so I am not suggesting a hard stop to the usage of all digital media in our churches. Far from it. I am simply stating the need for a shift away from our carefully manicured, perfectly professional digital presentations that come across as far too pristine and not nearly personal enough. The quality of the creativity and artistry is key. We must consistently invite the creatives and artists in our communities to lead the way and tell the story of God alongside us.

A WHEELCHAIR AND A MAT

Several years ago, my friend Roxanne attended a music festival called Outside Lands in San Francisco. During one of the performances, she heard a particularly loud uproar to her right. A group of people had noticed a young man in a wheelchair struggling to see the band over the mass of bodies standing above him. So these strangers turned their attention away from the stage and toward this young man. Several people knelt down. A small sea of hands tightly gripped tires and spokes and raised him high above the crowd. The young man, overcome by the emotion of the music and the moment, began to sing with arms lifted to the sky. The concert, which had simply been a performance up to that point, in an instant, became the soundtrack to something far more meaningful. And in this incredible act of grace, the entire experience changed. It was no longer about the music or the band or the performance. It was about a beautifully communal, participatory moment that would be remembered by countless people, leaving them marked and changed for the better. I wish I had been there. In person.

I can't help but think of the story in Luke 5, about the men who carry a paralyzed man up to the roof. Jesus is teaching in a house, it's packed, and there's no way to get this paralyzed man in to be healed. So the men carry him up to the roof and lower him down through the tiles to Jesus. Jesus forgives the man's sins and heals him. In Luke 5:20, we read that when Jesus sees their faith, he forgives the paralyzed man's sins and heals him. *Their* faith. Plural.

When we gather to sing, listen, and create together, we're reminded that sometimes we're the ones on the mat or in the wheelchair and sometimes we're the ones carrying others up to the roof or above the crowd. We see and hear one another. We smile and we greet and we pause from the digital distractions of our lives to remember that we are the people of God, and to embody that reality together. This is why it matters that we gather, in real time and in real space as real people. Because you can't lift a wheelchair or carry a mat or create worship or experience transcendence or be transformed over video.

PART 2

COMMUNITY

FOUR

REBUILDING BABEL

COMMUNITY IN THE DIGITAL AGE

To be everywhere is to be nowhere.

SENECA

A **FEW MONTHS AGO,** I was standing in my kitchen, phone in hand, editing a photo I'd recently taken of my kids. It was one of those all-timers, where both kids are happy and smiling, my infant son looking directly into the camera and my toddler daughter giving him a kiss on the cheek. A special moment caught in time that I wanted to remember and see over and over again. I decided to put just the right filter on it and crop it evenly so that it centered perfectly as my phone background. Per usual when staring at my phone, I was laser focused and tunnel vision set in . . . until I was interrupted. I felt a tug on my leg, immediately followed by these words: "No more email, Daddy."

My daughter had been standing there the whole time, watching and waiting for me to see her, lift her up in my arms, and play. But I was too distracted by the digital image of my daughter to see my actual daughter, waiting and disappointed. From her perspective, she thought I was checking email. So she did what she could to get me to see her. "No more email, Daddy." My heart sank. I put my phone down, picked her up in my arms, apologized, and we played.

In her book *Alone Together* Sherry Turkle writes, "Children have always competed for their parents' attention, but this generation has experienced something new. Previously, children had to deal with parents being off with work, friends, or each other. Today, children contend with parents who are physically close, tantalizingly so, but mentally elsewhere."[1] This is all too real in many of our experiences and not just between parents and their kids. A similarly dysfunctional dynamic is at work in a wide range of our relationships— with spouses, friends, classmates, coworkers, etc. We are often physically close—"tantalizingly so"—yet far from one another in the ways that truly matter.

Shortly after that experience with my daughter, I was having lunch alone. The restaurant was near a local high school that has an open campus policy, so shortly after I sat down to eat, several students began to file in together for a quick bite before heading back to class. Once again, I'd been on my phone—this time actually checking email. But when I saw the students walk in, I decided to people watch for a while, paying special attention to how they would interact while sharing a meal. What I saw saddened me but did not surprise me. In total, fourteen students ate at that restaurant during the lunch hour, all of them sitting in friend groups, not a single one of them alone. And in total, thirteen of them had a phone in their hands for the vast majority of the time, occasionally looking up to chat with one another, but for the most part, losing themselves to their digital content, all while sitting so tantalizingly close to other actual human beings. They were, in the words of Sherry Turkle's aptly titled book, "alone together." Entranced by the endless sea of digital possibilities, these kids were missing out on the very unique gift of analog presence surrounding them. While they were busy *communicating* with the digital world (many of them sending texts and Snapchat messages), they were squandering the opportunity to *commune* with the real people in their midst.

This is what community often looks like in the digital age. Lonely individuals falling prey, over and over again, to the great masquerade of digital technology—the ability to lull us into a state of isolation via the illusion of digital connection. In the next chapter, we'll get into the difference between *communicating* and *communing*. But first, how did we get here?

REBUILDING BABEL

In many ways, our technological ambitions have been getting the best of us since the very beginning. In Genesis 11, we find the Tower of Babel story. The entire world has one language and a common speech. They decide to build a city with a tower that will reach up to the heavens, in order to make a name for themselves and establish their greatness, in part, as a preventive measure to keep from being separated and losing their collective place.

But God thwarts their plans. He confuses their language and does the very thing they were so fearful of—he scatters them across the earth. Taken verse by verse, this story takes on an especially sobering and prophetic tone in the digital age.

"Now the whole world had one language and a common speech" (Genesis 11:1). The internet has brought us closer to this ancient reality than at any prior point in human history. Text messaging and video chat platforms allow us to instantaneously connect digitally with anyone around the world. Voice translation software allows for fairly seamless communication between people who don't speak the same language. Online connection has created what many call a singular "global village."

"As people moved eastward . . ." (Genesis 11:2). All of Genesis 1-11 shares one narrative arch, telling us the story of how humanity lost its way. At the fall, in Genesis 3, as the first humans rebel against God and sin enters the story, Adam and Eve are banished from Eden, the place God had designed for human flourishing. They move eastward to the new human reality of toil, labor, and pain. Eastward movement in this context, then, implies movement away from God's plan for human flourishing.

We see this same trajectory in the digital age. What was intended for our good and benefit is now leading us down a bleaker path. When it comes to the way we experience community, the very digital technologies that were meant to bring us together are now beginning to push us apart.

"They used brick instead of stone, and tar for mortar"
(Genesis 11:3). This is an explicit technological reference
and it's included in this ancient text for a reason. Stones are
naturally shaped over time, by their own mineral compo-
sition and surrounding geographic and atmospheric condi-
tions. Bricks are technologies created by manipulating
natural elements and shaping them for specified purposes.
They're also much better than stones for building towers
that reach the heavens. There's nothing inherently wrong
with bricks. Like all technologies, bricks can be used for good.
And like all technologies, bricks can be used to harm. As
we'll see in the next paragraph, this is what happens at
Babel. And it's what's happening in the digital age. The misuse
of digital technology is taking a helpful tool and making
it harmful.

"Come, let us build ourselves a city, with a tower that
reaches to the heavens, so that we may make a name for
ourselves" (Genesis 11:4). This tower, most likely an ancient
structure known as a ziggurat, is often misunderstood. Zig-
gurats in the ancient world were not structures designed to
create a pathway for people to get up to the heavens, but
rather, to create a pathway for the gods to come down to
people.[2] With this knowledge, at first glance, we may assume
that this is a good thing the people at Babel are doing, creating a structure to invite the presence of God among them.
But their motives are backward. Rather than longing for God
to dwell with them, this is an attempt at utilizing technology
for selfish gain. The people at Babel use technology for the
intent and purpose of "making a name" for themselves. The

Hebrew word for "name" here communicates fame, glory, and reputation. Sound familiar? On a macrolevel, this is an underlying distinctive of the Silicon Valley. On a microlevel, this is the engine that drives social media.

"So the LORD scattered them from there over all the earth, and they stopped building the city. That is why it was called Babel—because there the LORD confused the language of the whole world. From there the LORD scattered them over the face of the whole earth" (Genesis 11:8-9). God thwarts the people's selfish ambition at Babel by confusing their language and scattering them across the whole earth. Today, whether it's God directly intervening or it's the natural result of the digital age gone wrong (or some combination of both), this same cycle is clear:

UNITY ⟶ *TECHNOLOGY* ⟶ *SELFISH AMBITION* ⟶ *SCATTERING*

We are indeed more scattered than ever before. This is the ultimate paradox of the digital age: at the moment in human history when technology allows us to be more connected than ever, we are so very far apart, to the point that our very understanding of "community" has devolved into a sort of collection of isolated individuals.

SCATTERED, BABBLING, AND UNWISE

This scattering effect of the digital age has found its way into the life of the local church. Concerned conversations about declining attendance and engagement are commonplace among church leaders in America these days. If you're a pastor or church leader, chances are good that you've had these conversations.

Even the reasons people do participate in local churches reveal an important truth. According to a Pew Forum survey in 2017, the top three reasons people who attend church gave for their participation were:

1. *To become closer to God*

2. *So children will have a moral foundation*

3. *To make me a better person*[3]

All three are certainly good and valid reasons for participating in the life of the church. But there is something glaringly missing. There's no mention of the communal life of the church. No mention of the sense of belonging to something bigger than yourself. No mention of contributing to the greater kingdom good. Even in our churches, we gather as a scattered collection of individuals, with priority given to our own individual needs and desires, over and above the unique roles we might be called and equipped to play in the larger life of the body.

We live in an impatient, shallow, isolated culture. The idea of patiently journeying with a community of Jesus followers, doing the hard work of cultivating and excavating depth in our relationships with God and one another, and involving ourselves in the messy work of forging a meaningful community of diverse people doesn't seem like an attractive option. And the digital age is at the ready, offering a plethora of easier, quicker, shallower, more individualistic options.

In light of this, it's no surprise that church attendance and engagement as a whole is on the decline. According to a recent Barna study, only about one in three Americans has attended a church gathering in the past week, while two-thirds consider themselves unchurched or dechurched (those who were once a part of the church but no longer are).[4] In the Silicon Valley, where I live, those numbers rise dramatically, with nearly 90 percent of the population identifying as unchurched and/or dechurched. In short, fewer and fewer people are participating in the life of the local church. While there may be a wide range of factors when it comes to declining church participation, the influence of the digital age and the impact of its technologies play a significant role. As was the case at the Tower of Babel, our technological ambitions have moved us further away from one another and stolen from us the value of true community.

Several years ago, Facebook released an ad on television and online.[5] In the one-minute ad, a young woman is sitting at a dinner table with, presumably, her family. An older woman at the table, maybe her mother or an aunt, begins talking about her day and an experience she had at the supermarket. Noticeably bored, the young woman coyly pulls out her phone and proceeds to lose herself in Facebook notifications. First, a photo of a friend playing the drums. Then, another friend dancing in a ballet. Finally, some friends enjoying a day in the snow. As she "likes" each photo, they come to life all around her, the drumming, the dancing, and the falling snowflakes, drowning out her mother's (or aunt's or whoever's) humdrum monologue. The ad is well produced and compelling. Its magnetic pull is undeniable. Yes, when real life conversations get boring, I can just pull out my phone and plunge myself into more interesting things. I've given in to that temptation many times. But this ad highlights the sad and startling fiction of the digital age: that the best sort of communities and connections we can experience are the ones we can customize and craft to our own personal likings.

The church has been scarred by this pervasive lie. Many people want their local church to be customized and crafted to fit their specific needs, desires, and preferences. The concept of "church-shopping" could only exist in a culture like ours. We hop from church to church, with our detailed lists of exactly what we're looking for. To be sure, thoughtfully and prayerfully searching for the right church is a worthwhile pursuit. But when that search turns into a desperate hunt for our dream church, made up of just the right teaching, just the right music, just the right people and programs, something has gone terribly wrong. We've lost sense of what the church is actually supposed to be. As Dietrich Bonhoeffer once wrote, "Those who love their dream of a Christian community more than the Christian community itself become destroyers of that Christian community even though their personal intentions may be ever so honest, earnest, and sacrificial."[6]

This scattered search for a church that suits us just right has left many on the outside. The masses have simply given up on the search. The few who remain on the hunt find themselves increasingly frustrated. Blame is often cast in the form of blanket statements about what's wrong with the church as a whole, all the while missing the fact that the issue may very well be within—that what we are searching for may not exist and, more importantly, may not be at all what we need or even truly desire. But we hurl blame nonetheless. And in doing so, our scattered chaos leads to babbling.

In the Tower of Babel story, we read that "the LORD confused the language of the whole world" (Genesis 11:9). The same is true today. Because online connections are almost always immediate, quick, and truncated, digital dialogue leaves little to no room for reflection and nuance. So instead, we babble. We talk but we don't truly speak. We hear but we don't truly listen. In our babbling, we are quickly losing our ability to empathize, think deeply and critically, and offer the generosity of assuming the best of each other. Instead, we quip and squabble, as quickly and loudly as we can. For an example, find and read any Twitter thread of individuals arguing about any given thing.

One of the most dangerous things about our babbling is its propensity to radically lower our expectations of one another. As our interactions are mostly comprised of our shallowest articulations, we begin to see others as only superficial caricatures. This in turn has a toxic effect on our understanding of community. The idea of gathering and journeying with others for the long haul loses its value and meaning. This is one of the most contradictory byproducts of the digital age, that our easy-access connections have actually made us so adept at pushing each other away, sequestering ourselves from any real connection. Sherry Turkle astutely points out that this dynamic increases "the risk that we come to see others as objects to be accessed—and only for the parts we find useful, comforting, or amusing."[7] Because we see one another simply as "objects to be accessed" for our own personal benefit, when others are not "useful, comforting, or amusing" we become incredulous at the fact that they are not faithfully orbiting around our own personal needs and desires. Nothing could be more detrimental to community—especially the sort of community that the church is called to be.

Turkle has observed that on average it takes about seven minutes for a conversation to really get going. The first few minutes are often shallow, disjointed, and even boring.[8] But in order to achieve any sort of depth and meaning in our conversations and interactions with others, in order to create a sense of true community and connection, we must remember that "it is often in the moments when we stumble and hesitate and fall silent that we reveal ourselves to each other."[9]

But in the digital age there's no space for that. We do not gift one another the allowance of being able to stumble, hesitate, and fall from time to time. We do not afford one another the grace of being helped back up, to regather our thoughts, reconsider our ideas, and reengage in the dialogue. As a result, in order to protect ourselves, we have become clever. We've become skilled at making fast-paced, quick-hit points and winning arguments. The problem is, meaningful community is forged slowly, over time, with much compromise and understanding. Healthy communities invite nuance and emphasize reflective responses over rash reactions. Real connections within real communities are realized only as we walk together down the path of wisdom, not cleverness. And in the digital age, there is a tremendous wisdom gap. We have all the information we need to be clever. What we don't have is the commitment to journeying alongside others up the narrow and often difficult path of wisdom. As James Emery White puts it, we face "a widening chasm between wisdom and information."[10]

This is where the church can and must step in, to offer a new vision for what community can and should be—an alternative, transcendent space where unlikely people gather to listen and speak, to reflect and respond, to journey together for the long haul down the path of wisdom. But instead, so many of our churches have gone the other way.

CHURCH ONLINE OR
AN ONLINE CHURCH?

As digital technology ramps up the ability to create and scale online presence, churches have run headlong into these new opportunities for growth and expanding reach. There are entire companies whose sole purpose is to help churches create not only an online presence but an entire "online church." One of the leading companies describes their technology this way: "It's more than a video player. It's an agreeing-in-prayer, real-life, around-the-world community builder. With Chat, Live Prayer, and synced video streaming, [we're] all about doing church together."[11] The assumptions are clear—through digital technology and online presence, we can build community; with tools like real-time text and video chats, all of the standard church-community components can be re-created—conversation, prayer, sharing of content—and this allows us to "do church together."

One of the churches that first pioneered the online movement is now one of the largest "online churches" in the world, with more than three hundred thousand people watching their content every week. On their webpage, in a brief video explaining the story behind their church, the pastor says, "We thought, what if we could use technology to create an environment online that would bring and gather people together, to have church and to be a church that existed and met online?"[12] Again, there is another set of clear assumptions—"gathering," "coming together," and "being a church" are all realities that can be experienced online, without actual human presence or physical proximity.

According to the company Livestream, in 2016 more than twenty-seven hundred churches used their live video streaming platform to broadcast more than one hundred sixty-six thousand events online.[13] This is just one company's data. When we factor in the variety of other platforms, including giants such as Facebook and Instagram (both of which offer live streaming now), it's safe to assume that the total number of churches streaming worship gatherings and other content is astronomical. Even the Amish have an online presence these days![14] Consistently, what all of these online platforms emphasize as their primary means of creating "community" is the ability to have live chats, to discuss the viewing content, and to ask for and receive prayer in real time. These features have the potential to be helpful and even transformative when appropriately understood and responsibly utilized. I have heard several stories of people living in parts of the world where Christianity is a punishable offense, coming to put their faith in Jesus Christ because of sermons they watched in secret and follow-up chats with Christians via these sorts of online platforms. This could and would not have happened without digital technology and Christians leveraging their ingenuity to share the gospel in creative ways. But if we were able to ask these brave men and women in these dangerous parts of the world what they long for now that they are Christians, it's not a reach to assume that they would mention some form of actual, physical, present community—to be able to gather with other Christians, in real time and real space, to be encouraged, challenged, inspired, and to not only know they weren't alone but also to experience that directly. This much is made clear in the courageous ways that persecuted Christians continue to physically gather in underground churches in various parts of the world, often at the risk of imprisonment and even death.

In America, we have an altogether different problem. This same technology is being used not primarily to reach unreached people groups and those who would otherwise never hear the good news of Jesus. Instead, it's primarily being used to saturate the Christian church marketplace, providing a wider variety of high-end (and sometimes not-so-high-end) products for the church-shopping masses. A leader at a church that regularly uses Livestream to do "online church" writes this endorsement of the online platform service: "We were looking for a way to get our product out there and this was the easiest to access for a wide variety of people and more user-friendly."[15]

"Get our product out there."

"Easy to access."

"User-friendly."

This is not the language of community; it is the language of commodity. An "online church" is more a *product* to be consumed than it is a *people* to be joined. Community isn't about *getting a product out there* but about *gathering people wherever they are.* Yet, so many of our churches continue to push into online spaces and call it community and connection. And in doing so, we are doing tremendous damage to the very communities and connections we so desperately long to see. Sherry Turkle writes, "When online life becomes your game, there are new complications. If lonely, you can find continual connection. But this may leave you more isolated, without real people around you. So you may return to the Internet for another hit of what feels like connection."[16] The more our churches promote and push people toward online spaces for community and connection, the more lonely and isolated they will be in the end, as they continue returning to these online spaces to fulfill what they never actually could.

I recently saw a video that epitomizes this entire online approach to community. It was tweeted by Judah Smith, a nationally recognized church leader. I don't know him personally but I've been inspired by much of his ministry and admire him for much of the work he's done. I assume that at his core, his desire is to make Jesus known to as many as possible. In the video, he announced that his church was launching a new location. I was surprised to hear him say this:

We have a new location. And that location is *everywhere.* That's right. It's global. . . . You can go to your favorite app store; you can download it today. There's gonna be daily content, fresh content, brand new content every single day. One of my favorite parts is before the live service begins, you can meet new people in the 'lobby'; I know that sounds crazy but there's an actual lobby where you can connect, meet people. We're passionate about connecting people with God and each other and this is going to be maybe the most effective platform we have ever used in doing so, where people can actually build real tactile relationships all over the world.[17]

There are two fundamental problems with Smith's announcement. First, the primary focus of this new church "location" is content. As he reiterates, "fresh content, daily content, brand new content." But church communities cannot be built primarily around content, because while great content can inform and even inspire, content alone is never enough to transform us. Transformation in the life of the church is always an analog experience, as we journey shoulder to shoulder with other people, gathering in real ways as real people, to invite God to change us individually and collectively. We experience this transformation in a variety of ways—singing together, listening and speaking of God's grace and truth, the breaking of bread, the sharing of resources, the giving of our time and energy and creativity, remembering and celebrating via the sacraments, and on and on—but all of these ways are in some form or fashion, tangible and physical. Content matters inasmuch as it moves us toward real participation and action within the actual church community. Smith seems to realize this and proceeds to declare that there is a "lobby" where you can connect and meet people. This leads to the second problem.

As much as he'd like to say that this new location will have an "actual lobby," the fact remains, it doesn't. And it can't do what is promised—build "real tactile relationships"—because "tactile" by its definition demands in-person presence.

I am on an ongoing text thread with several close friends. We text almost daily and most of the time it's just lighthearted chatter. But from time to time, we share more deeply. It might be a challenge we're facing or discouragement we're fighting or a question we're wrestling. When this happens, we text prayers and encouragement to one another. I've been affected personally by these texts on countless occasions. But my best times with these guys are the couple of times a month we get together, in person, to share a meal at our go-to local Thai place. Over curry and crab rangoons, we experience incarnation—the words and ideas and truncated conversations on our text thread come to life in real, present, viscerally human ways. Our "LOLs" become actual, quite loud laughter. This is what our digital and online connections do at their best—they push and prod us forward, increasing our hunger and desire for the real thing.

I am not suggesting we do away with online presence altogether. Far from it. As I wrote earlier, when understood appropriately and utilized responsibly, digital technology affords us brand new opportunities to share the gospel, as well as to encourage and challenge one another. As Ed Stetzer puts it, "A church should be online, but I don't think it should be an online church."[18] What I am suggesting is that we understand and utilize online platforms for what they truly are—a helpful digital means to a greater incarnational end.

EKKLĒSIA

Ekklēsia is the Greek word that's most often translated "church." It's found over one hundred times in the New Testament. When people hear the word "church" today, they typically understand it in a few ways. Some people think of a specific building in a specific town, maybe a place they attend from time to time. Other people think of a national or global religious institution and its systematic presence in the world. Still others think of the two-millennia-old historical entity of the Christian church. But in the Bible, the word *ekklēsia* refers to a local community of Christ followers a predominant majority of the time. *Ekklēsia*, or "church," in a biblical sense, is almost always a group of people who gathered regularly to worship, share their lives with one another, and learn and live the way of Jesus together. The word itself is a compound of two Greek words meaning "called" (*kaleō*) and "from" or "out of" (*ek*); in broader terms, the word was understood to mean, "an assembly or gathering of citizens called out from their homes into some public place."

The New Testament, written specifically to and about these local churches, is full of very specific instructions on how to live and be as this called out community of God's people:

"Serve one another humbly in love" (Galatians 5:13).

"Be completely humble and gentle; be patient, bearing with one another in love" (Ephesians 4:2).

"[Speak] to one another with psalms, hymns, and songs from the Spirit. Sing and make music from your heart to the Lord" (Ephesians 5:19).

"Let the message of Christ dwell among you richly as you teach and admonish one another with all wisdom through psalms, hymns, and songs from the Spirit, singing to God with gratitude in your hearts" (Colossians 3:16).

"Spur one another on toward love and good deeds, not giving up meeting together, as some are in the habit of doing, but encouraging one another" (Hebrews 10:24-25).

"Offer hospitality to one another without grumbling" (1 Peter 4:9).

"But if we walk in the light, as he is in the light, we have fellowship with one another" (1 John 1:7).

"Therefore confess your sins to each other and pray for each other" (James 5:16).

"They devoted themselves to the apostles' teaching and to fellowship, to the breaking of bread and to prayer" (Acts 2:42).

These are not outdated characteristics of the ancient church. These are the markers of the church today and for all time.

Serve one another.

Bear with one another in love.

Speak and sing the words of God together.

Make music together.

Teach and challenge one another.

Keep one another accountable.

Spur one another on toward love and good deeds.

Do not give up meeting together.

Be hospitable to one another.

Experience fellowship together.

Confess to one another.

Pray for one another.

Eat and drink together.

All these things are difficult at best, and impossible at worst, to do online. These practices of the church, the gathered community of God's people, require physical presence. Community always has been and always will be an analog experience. So what can and must the church do in the digital age to fight the technological tide of online connections and invite people into these analog community experiences?

FIVE

A TAX COLLECTOR AND A ZEALOT WALK INTO A CROSSFIT

ANALOG COMMUNITY

Commitment matters more than compatibility.

BRETT McCRACKEN

MY WIFE AND I met at church. I was an intern in the youth ministry when she first visited. We became fast friends and eventually, after much cajoling on my part, we started dating. About a year into our relationship, I left that church to become the youth pastor at another church in town. But Jenny was serving as a volunteer middle school small group leader and was committed to being with her girls until they graduated. So she stayed put.

During that year at different churches, Jenny decided to get baptized. I said I'd be there, of course. But as it turns out, I wasn't. I hadn't looked at my calendar carefully and later realized that I had a scheduling conflict. Additionally, the thought of seeing old faces, making small talk, and catching up sounded exhausting to me at the time. I decided not to go. She was understandably disappointed but I tried to justify it by saying that there'd be someone there taking video of the baptism and that I'd watch it later.

I have significant regret about missing such a pivotal moment in Jenny's journey with Christ. It's an ongoing source of embarrassment for me. To miss such an important day and to offer the justification that I would watch it on video later is laughable. It was a selfish choice guided by my penchant for convenience. When it comes to the truly important stuff of life, there's no substitute for showing up. As inconvenient as it is sometimes, lending our whole-bodied presence to others matters a great deal. But in the digital age, we are being told otherwise.

A few years ago, Mark Zuckerberg, the founder of Facebook, delivered a speech in Chicago in which he presented Facebook's plans to broaden their platform's ability to connect people with one another. "If we can do this," he said, "it will not only turn around the whole decline in community membership we've seen for decades, it will start to strengthen our social fabric and bring the world closer together."[1] He went on to claim that in light of recent declines, Facebook was becoming "the new church."[2]

Is Zuckerberg right? Can we strengthen our social fabric and bring the world closer together through online platforms?

Is the need for real spaces where real people can rub shoulders and share life going away in the digital age? Is Facebook the new church?

In a compelling response to Zuckerberg, journalist Peter Ormerod begs to differ. He writes this:

> At their best, churches offer a perspective on life funda-mentally opposed to the culture Facebook encourages and upon which it feeds. . . . Churches, at their best, bring us into contact with people we would never think of as friends. There are cliques, of course. But we all come to the same table and drink from the same cup and sing the same songs and say the same prayers. The Lord's Prayer, after all, is not in the singular, but the plural: "Give us today our daily bread." It's a breaking down of barriers, an awareness of mutual responsibility and dependence, a celebration of brokenness. It's an unsanitized expe-rience of humanity, and all the healthier for it. . . . A good church is more than just a social network: it's a place of transcendence, space, silence, peace, devotion, richness and depth. No matter how grand Zuckerberg's visions may be, they will never compete.[3]

The church is designed to do and to be something Facebook, or any other social media or online platform, could never be—a real gathering of real people, as unlikely and different as they are. As Scot McKnight writes, "The church is God's grand experiment, in which differents get connected, unlikes form a fellowship, and the formerly segregated are integrated. They are to be one."[4] The church is what the ever-customizable, carefully curated and personalized world of social media and online platforms could never be. The church is community built primarily on commitment, over and above compatibility or comfort. The church is family.

THE CHURCH AS FAMILY

Modern evangelical Christianity often presents an incomplete-at-best and distorted-at-worst version of what it means to be a follower of Jesus. Due in part to the centuries-long rise of individualism in society at large, many Christians today have been taught, both implicitly and explicitly, that salvation in Christ is an individual experience.

But the Bible begs to differ. Jesus and the New Testament writers make clear that those who profess Christ as Lord and Savior are not simply saved as isolated individuals but, most importantly, are saved *into* the community of God. As Joseph Hellerman writes, "In Scripture, salvation is a community-creating event."[5] And this isn't just any sort of community—it's family.

Because we've so often read and heard the idea that Christians are "brothers and sisters in Christ," we've lost sense of the radical reorientation that this idea commands. In the first-century world of the New Testament, the most important societal bond was between siblings. In today's culture, we often think of marriage as the relationship that holds our primary allegiance. But this was not so during the time of Jesus. In the patriarchal system of the day, it was the blood relationship between siblings who shared the same father that was the most significant familial bond, over and above the contractual bond between spouses.[6] In light of this, the fact that the New Testament writers address followers of Jesus as "brothers and sisters" more than a hundred times takes on added significance. Salvation draws us into a brand-new family, one that replaces our previously held allegiances. In Christ, we become children of our one Father God, siblings bound by the blood of Jesus himself.

Our family experiences vary broadly, which makes the idea of the church as a family challenging. For some, family gatherings are highly anticipated and when they're over, we can't wait to gather again. For others, family gatherings bring a sense of dread and we scramble for reasonable excuses to get out of attending. I grew up an only child with a single mom, and the dynamics of siblings and large families were foreign to me. But with two young children now, I'm learning through observation just how up and down sibling relationships can be. One minute, it's joy and laughter. The next minute, it's anger and screaming.

Their relationship is volatile and often unpredictable. But their kinship is not. Their kinship is undeniable and unceasing. They will forever be brother and sister. Because of this connection between them, which they did not choose, the only choice they have now is how they will navigate the bounded reality of their kinship.

So it is with the family of God, the church. We are bound to one another. We did not *choose* it; we were *saved* into it. The only choice we have is how we will care for and cultivate this kinship.

One of the key values of the digital age is *individualism*, which leads to *isolation* when taken to its extreme. The more we present "church" as an easily accessible product to be consumed digitally, when it's preferable and convenient for us individually, the more we ground people in the misunderstanding that the Christian life is a solo venture, customizable according to our personal preferences. But this solo approach to following Jesus invariably moves us away from the call to Christian kinship. It is not enough to simply declare with our mouths, "We're brothers and sisters in Christ!" Since the beginning, this new family-of-God dynamic has been intended to disrupt our real, actual lives.

In his first letter to the Corinthian church, Paul reprimands the Christians there for taking one another to court.[7] This was a common way of settling disputes at the time—an acceptable, socially normative way to behave. The only situation in which taking someone to court was frowned upon was when siblings would attempt to do so with one another. Because the sibling relationship was the primary social bond in society at the time, it was considered inappropriate for brothers and sisters to take legal action against each other. Their disputes were to remain a family matter, settled personally.

Writing to the Christians in Corinth, who were physically unrelated to one another, Paul says, "The very fact that you have lawsuits among you means you have been completely defeated already. Why not rather be wronged? Why not rather be cheated? Instead, you yourselves cheat and do wrong, and you do this to your brothers and sisters" (1 Corinthians 6:7-8). Paul addresses these previously unrelated individuals as "brothers and sisters," bound up together in the family of God through Christ, and he applies this new kinship to their real, actual lives. No longer are the Corinthian Christians to take one another to court to settle disputes. Such a thing is unacceptable between brothers and sisters. And that is what they are now, truly. Their kinship is not just a theological idea or poetic metaphor—it is their new everyday reality, disrupting their real, actual lives.

In the same way, being part of the family of God ought to disrupt our real, actual lives today. And one of the most needed disruptions to our digitally oversaturated lives is the call to go offline and actually show up—as healthy families do.

COMMUNE OR COMMUNICATE

Digital communities are convenient and customizable. They are based on our preferences and designed to be easily and quickly chosen or unchosen. Don't like something someone said on your Facebook feed? Unfriend them. Annoyed by someone's endless stream of gratuitous food pics on your Instagram feed? Unfollow them. Irritated by the opinions of someone on Twitter? Block them. All these options can be activated with the push of a button in a split second.

But analog communities are different. When we show up, in the flesh, it's not as easy to unfriend, unfollow, and block. Because despite our differences and incompatibility, *here we are*. Despite our disconnections and often divergent perspectives on things, we've gathered together and committed to giving a portion of our lives and energies to a particular people in a particular place at a particular time. This is what families look like. Like it or not, we are connected.

This is a risky decision, to be sure. Analog communities are based not on *preferences* but on *presence*. They are formed simply by who's there, and it's much more difficult to choose or unchoose them. (Certainly, even with analog communities, choice is involved. In the consumerist landscape of churches today, people regularly drive right past several churches to go to one that better suits their preferences. There is much to say about this, but for our purposes here, we'll focus on the initial choice to be made—the choice to show up at all). As antithetical as it may seem, these are exactly the sorts of spaces we need in order to more fully become the people of God together. This is why community in the context of the church must be analog.

Earlier, I mentioned the chasm between *communicating* and *communing*. The digital age has amplified our ability to *communicate* at the tragic cost of our aptitude for *communing*. It's important to note here that these words each have an elasticity of meaning and the differences between them can easily become muddled, at least on technical and linguistic levels. The goal is not to debate their etymology but to differentiate their core ideas.

To *communicate* is primarily about the exchange of *information*.

To *commune* is primarily about the exchange of *presence*.

To be sure, *communicating* isn't always simple and it isn't always easy. Miscommunication happens all the time. But to *commune* is the more difficult task. It requires more of us: more of our attention, empathy, and compassion. In light of this, why must community within the context of the church be analog? Because while we can certainly *communicate* digitally, we can only *commune* in analog.

Digital technologies are exceptional and efficient when it comes to the exchange of information, but they are abject failures when it comes to the exchange of presence. We've all had the experience of receiving an email or a text that we completely misunderstood because, while the information relayed digitally was clear enough, the lack of body language, tone of voice, posture, and so on left us in the dark regarding the full expression of the message received. There is a time and place for communication. But what human beings need most is the regular, ongoing experience of communing with other human beings.

And the church is meant to commune. That is her intrinsic design and intent. This means that things will often get messy, complicated, and inconvenient in our church communities— but that's exactly how it's supposed to be. In working out and working through the mess, complications, and inconveniences together, we are transformed into the people God is calling us to be.

Like getting physically fit or learning a new skill, our discomfort is itself the very sign that we are making progress. And in turn, when our community is comfortable, as is almost always the case when experienced at a digital arm's length, it is a surefire sign of stagnation. As Brett McCracken writes in his book *Uncomfortable*, "There is a reverse correlation between the comfortability of Christianity and its vibrancy. When the Christian church is comfortable and cultural, she tends to be weak. When she is uncomfortable and countercultural, she tends to be strong."[8] This is not a new phenomenon. Since its earliest days, the church has been about unlikely people gathering as family, in spite of their differences, living in uncomfortable community with one another, learning together to become one in the transformative presence and by the transformative power of Jesus Christ.

YET, HERE WE ARE

In Matthew's Gospel, there is a brief description of the twelve apostles, Jesus' closest friends during his earthly ministry:

> These are the names of the twelve apostles: first, Simon (who is called Peter) and his brother Andrew; James son of Zebedee, and his brother John; Philip and Bartholomew; Thomas and Matthew the tax collector; James son of Alphaeus, and Thaddaeus; Simon the Zealot and Judas Iscariot, who betrayed him. (Matthew 10:2-4)

Only two characters have unique descriptors attached to their names: Matthew, who is noted as a tax collector, and Simon, who is noted as a zealot. These descriptors seem trivial to us today, but in the first-century Jewish world they were shocking. As a tax collector, Matthew was a Jewish man employed by the Roman government. Tax collectors were almost always corrupt, allowed by Rome to skim off the top as they collected taxes from their own people on behalf of the oppressive ruling empire. Tax collectors were seen and treated as traitors who were getting rich by stealing from their own people under the authority of the enemy. Zealots stood on the furthest opposite end of the sociopolitical spectrum. They were fanatical, anti-Rome activists who used violent means in their attempts to overthrow their oppressors. They didn't simply attack Roman soldiers; some of their favorite targets were tax collectors, being the traitors that they were.

Jesus, in handpicking the young men who would follow him and eventually be tasked with launching the Christian movement, intentionally includes both a tax collector and a zealot. Imagine the dissonance and resentment between them. Imagine the primal urges both men must have felt from time to time. Simon very well may have needed to suppress his desire to pounce on Matthew and pummel him with anything he could get his hands on. Matthew very well may have existed in a constant state of anxiety and fear, looking over his shoulder, wondering where Simon was at all times, afraid of bloody retribution coming his way. And yet, we read nothing of that in the Gospels. It doesn't mean the tension didn't exist. Almost certainly it did, especially early on in their years together with Jesus. But what it also means is that somehow, some way, these two diametrically opposed individuals experienced, and at some point, embraced one another as brothers brought together by a force much stronger than any of their own past histories, experiences, and opinions of one another.

At some point, Simon and Matthew must have looked at each other and thought, *I did not choose you. You did not choose me. Yet, here we are.*

This beautiful ethic of the Jesus way of life is amplified in later writings. The apostle Paul, himself once a persecutor of the Christians, writes, "Live in harmony with one another.

Do not be proud, but be willing to associate with people of low position. Do not be conceited" (Romans 12:16). Peter, who had been with Jesus alongside Matthew and Simon, writes, "All of you, be like-minded, be sympathetic, love one another, be compassionate and humble" (1 Peter 3:8).

Because it's so easy to unfriend, unfollow, and block online, it is extremely difficult, if not downright impossible, for us to maintain this type of sustained presence with those unlike us in digital communities. And without the experience of a sustained journey alongside those unlike us, learning their stories and perspectives, we can very easily spiral into the black holes of our own carefully self-curated (and now, machine algorithm-curated) digital worlds. This is a dangerous proposition. The writer Jaron Lanier notes that in the digital age, "when we're all seeing different, private worlds, then our cues to one another become meaningless. Our perception of actual reality . . . suffers."[9] Yet, these "private worlds" of our own digital making are the places so many people spend most of their waking hours.

And so, the sort of community Jesus brought together during his earthly ministry was emphatically contradictory to the sorts of communities that thrive in the digital age. Preferences, shared interests, personality quirks, personal histories, even worldviews and philosophies are set aside and replaced by a set of common goals: to follow Jesus together, to learn, embody, and live out his ways, to join him in the work of inviting people into the kingdom of God. And this same approach to community is upheld and expanded by the earliest leaders of the Christian movement and the churches that formed under their leadership. The church, at her best, continues to invite people into these same sorts of communities today; and in doing so, we display for the world a transcendent vision of a new way to belong. Now, possibly more than ever, we must help our churches craft and cultivate this transcendent vision of community because there is both a greater need and a greater opportunity than ever before.

The need exists because the digital age has disconnected and detached us from one another in ways completely unique to our current moment in history. True analog community is what the world is hungry for, whether they know it or not.

The opportunity exists because the church is poised to offer opportunities to experience this sort of community in ways that no other thing on earth is. There are plenty of affinity group opportunities all around us, but only the church can be a place where even our affinities are set aside to gather around the profoundly good news that God loves us, is with us, and will make all wrong things right someday.

Right before his arrest and subsequent death on the cross, Jesus prays alone in the garden. He prays for us, all those who would come to put their faith in him through the message of his disciples and the churches they would eventually launch. He says this:

> My prayer is not for them alone. I pray also for those who will believe in me through their message, that all of them may be one, Father, just as you are in me and I am in you. May they also be in us so that the world may believe that you have sent me. I have given them the

glory that you gave me, that they may be one as we are one—I in them and you in me—so that they may be brought to complete unity. Then the world will know that you sent me and have loved them even as you have loved me. (John 17:20-23)

Jesus prays that we would be brought to complete unity. All of us.

The tax collectors and the zealots.
The conservatives and the liberals.
The majority and the minorities.
The healthy and the sick.
The rich and the poor.
The haves and the have-nots.
The men and the women.
The young and the old.
The joyful and the mournful.
The hopeful and the heartbroken.

All of us brought together into complete unity, because then "the world will know." When we gather in real time and in real space, despite our differences, we proclaim to the world that there is a God who loves us and sent his Son to change everything. Simultaneously, we experience a brand-new connection between us—Jesus Christ himself. As Dietrich Bonhoeffer writes, "The more genuine and the deeper our community becomes, the more everything else between us will recede, and the more clearly and purely will Jesus Christ and his work become the one and only thing that is alive between us. We have one another only through Christ, but through Christ we really do have one another."[10]

There are many ways to do this, but let me share a few simple ideas.

THE ART OF GATHERING

CrossFit is a worldwide physical-fitness phenomenon. Since 2005, CrossFit has grown from just thirteen gyms to more than thirteen thousand gyms globally and four million active participants today.[11] I have many friends who are avid CrossFit devotees and all of them claim that the most beneficial part of being involved has to do with the community they experience. One of these friends, Justin, is the general manager of the CrossFit Games and helped negotiate the deal to have the Games broadcast on national television. He says that "the sense of community is what sets CrossFit apart. A gym where there are no TVs or mirrors. People leave their cell phones in the car and there are no headphones allowed. Eye contact, conversation, and encouragement are the norm. The friendships formed when you are exhausted are some of the strongest. And those relationships extend beyond the walls of the gym. Try as you like, you can't replicate that online."

Justin is exactly right. That sort of community can't be replicated online. I've tried. Over the years, I've intermittently gotten into a few different online workout videos. I turn on YouTube and follow along as the instructor guides me through the routines. It's helpful for a while but something about the experience lacks the motivating factor necessary to stay with it. There's no accountability, and simply put, it's lonely. This is why going to a gym, working out alongside others, and receiving encouragement and challenge will always outpace even the very best online programs we may find. This is also why the most physically fit people you know typically have some sort of workout community around them, be it CrossFit, another gym, a running group, a rec-league team, or something else. Gathering matters. Being shoulder to shoulder and blocking out time in our busy schedules to focus on a particular goal—alongside others who share the same goal—keeps us motivated, encouraged, challenged, and leads to transformation. None of this can be replicated online.

In the digital age, we're in danger of losing the art of gathering in our churches. In the name of relevance and in our attempts to "meet people where they are," many churches are recklessly leaning into online platforms to bring us together. But this strong focus on digital presence, when it is not tempered and put in its rightful place, can have an adverse effect on the sorts of analog community experiences that people truly need.

As we serve and lead our churches, we must be thoughtful and creative in designing gathering spaces and opportunities that invite people to put down their screens and show up with their whole selves. The church must recapture the art of gathering by inviting people in a compelling way to gather as real people in real places in real time, in order to experience real transformation.

PRESENCE IN PAIN

Todd and Natalie were dear friends of mine at a church plant I was a part of several years ago. They'd met in college and were engaged shortly after graduation. In the months leading up to their wedding day, Natalie was diagnosed with an aggressive form of stomach cancer and given a very short time to live. When I met them they were newlyweds. They became an inspiration to many in our church community as they remained faithful to Jesus and to one another despite the cloud of pain and loss looming overhead.

Late one night, Todd called and asked me to come over. Natalie was in her final hours. I rushed to their home as quickly as I could, battling tears and fearful of not knowing what to say or do. I arrived and sat at Natalie's bedside. She'd fallen asleep but was still breathing, just barely. I held her hand, read Scripture, prayed, and cried. I hugged Todd. I didn't say much, as words seemed inappropriate in the moment. I gave Todd nothing more than my presence in the midst of his pain. It didn't feel like it was enough, but it was all I had, so it's what I gave. Natalie passed away a few hours later.

That late night with Todd and Natalie was painful, but it was also one of the most profound experiences of community I've ever had. To be present in the midst of pain has a bonding effect that is impossible to fully replicate digitally. Our churches must become places where people can shut off the incessant pleas of the digital world to show off their best selves, to highlight their highlights, to self-promote with facades of happiness and fun, and instead, to show up with all of their pain, knowing that they will be seen, embraced, and loved. Dietrich Bonhoeffer wrote, "Every Christian community must know that not only do the weak need the strong, but also that the strong cannot exist without the weak. The elimination of the weak is the death of the community."[12] One of the reasons this is so crucially important is that the line between weak and strong is not a line between me and you, or us and them; it's a line down the center of each and every one of us. Sometimes I'm weak and sometimes I'm strong. So are you. And the beauty of an analog community that invites us in whether we are weak or strong is that we are surrounded and supported by those who may be strong when we are weak but know exactly what it means to be in our shoes, and vice versa.

Every church where I've served has had a prayer room or a prayer area in the worship gatherings. This is one of the most effective ways to create an analog community where people are invited to be present in one another's pain. Just about every Sunday at our church, during the closing set of musical worship, we invite people to join us in the prayer room. We have a dedicated and trained team there, ready to listen, pray, and be present in the midst of pain. The room is rarely empty. There are always people who go to seek prayer, comfort, and presence. In all churches, we must create, emphasize, and consistently invite people into these sorts of spaces, not simply as an aside but as one of the primary reasons we gather.

Several churches I know have created teams of what some would call "neighborhood ministers." These men and women, sometimes paid staff and sometimes lay leaders, serve by pastorally overseeing and ministering to specific neighborhoods in the church's city. They meet regularly with the small group leaders whose groups gather in those neighborhoods. They do home, work, and hospital visitations. They pray for the needs of their people, raise awareness of these needs, and rally others in the same neighborhood to adequately meet those needs. This focused pastoral presence for different segments of the community helps to create a sense of belonging that can be difficult to achieve in the larger worship gatherings (especially at larger churches).

In his book *Tribe*, Sebastian Junger writes about instances throughout history when people surprisingly rallied together in the midst of unspeakable tragedy. Earthquakes, tsunamis, and wars that led to the loss of countless lives resulted in the shocking rise of emboldened communities that began to care and provide for one another, friends and strangers alike, in ways never before seen. He writes, "What catastrophes seem to do—sometimes in the span of a few minutes—is turn back the clock on ten thousand years of social evolution. Self-interest gets subsumed into group interest because there is no survival outside group survival, and that creates a social bond that many people sorely miss."[13] When we create spaces in our churches for people to be present with one another in the midst of their pain, we too create opportunities for these social bonds. But what they sorely miss is so much more than social. It's spiritual. It's the way God designed us to be human with one another, bringing all of ourselves, all our pain, the weak and the strong, gathering as one.

SPARE THE GOATS BUT DON'T STOP CONFESSING

Jenny and I had an argument recently. We both said hurtful things to one another. After sleeping on it and a quiet morning, we'd calmed down and went on with our days, both heading off to work in different directions. Hours later, while in the office, I texted my wife two simple words: "I'm sorry."

Later that evening we both arrived home and Jenny said something surprising to me: "You still haven't said you're sorry." I thought, *What do you mean? I texted you.* But she was right. I'd texted it but I hadn't said it. So I put my phone down and said the words I'd texted earlier, "I'm sorry." It didn't end there though. She said she was sorry too. Then we talked. We discussed, face to face, what had gone wrong, why we each reacted the way we did, and how we could both get better together. The text was an easy *apology*. But speaking the words in person allowed for a much more involved conversation, which eventually led to *confession*. The difference here is crucially important.

In our culture, *apologies* are what we're most familiar with. We wrong someone. We recognize the wrong. We apologize. Hopefully they forgive. Maybe we forget. Done. Simple enough. But in the Bible, when people commit wrongdoing (read: sin) there is always a much more involved, public, and communal response. A central part of this response is *confession*. For example, in the book of Leviticus, God gives specific instructions to Moses regarding how the people of Israel are to maintain a right relationship with him, in spite of their consistent wrongdoing. God tells Moses that his brother Aaron "shall bring forth the live goat. He is to lay both hands on the head of the live goat and *confess* over it all the wickedness and rebellion of the Israelites—all their sins—and put them on the goat's head. He shall send the goat away into the wilderness in the care of someone appointed for the task. The goat will carry on itself all their sins to a remote place; and the man shall release it in the wilderness" (Leviticus 16: 20-22, emphasis mine). Aaron, as the first high priest of Israel, would publicly and collectively confess *all the sins* of the nation and symbolically place them on the head of this goat. This goat, known as the scapegoat, was then sent off into the wilderness to die, carrying to its grave the confessed sins of the people. Much more involved, public, and communal than a simple "I'm sorry" text.

Some may argue that because of Christ's atoning work on the cross, which took the place of the scapegoat, as well as a countless array of (usually bloody) confessional practices we find in the Old Testament, Christians no longer need to think of confession in the same way. While I certainly agree that we can spare the goats now, the New Testament does continue to press the involved, public, and communal nature of Christian confession. The apostle James instructs us, "Therefore confess your sins to each other and pray for each other so that you may be healed" (James 5:16). This is more than a private prayer seeking forgiveness from God or a convenient apology via text or email. This is confession much more in line with the experiences of the Israelites during the time of Moses, as they gathered to admit their wrongdoing, to commit to living more faithfully moving forward, and to rejoice collectively in their newfound forgiveness and freedom. In fact, both the Hebrew (*yadah*) and Greek (*exomologeō*) words for confession also mean "to praise" and "to give thanks joyfully." In confession, there is both remorse and rejoicing; there is both conviction and commitment. It is a much richer and more meaningful act than the simple, one-sided apologies that have become so absurdly easy via text or email in the digital age.

In the digital age, one of the most transformative things we can do in our churches is create space and opportunity for confession. As awkward and uneasy as it may seem, taking time to admit our wrongdoing toward God and others, together, is a powerful and much-needed practice. Ultimately, without Christian confession, we cannot fully experience and enter into the life of the community. Dietrich Bonhoeffer explains it this way: "In confession there takes place a breakthrough to community. Sin wants to be alone with people. It takes them away from the community. The lonelier people become, the more destructive the power of sin over them. The more deeply they become entangled in it, the more unholy is their loneliness. Sin wants to remain unknown. It shuns the light."[14]

Sin is a beast that thrives in the deep seas of isolation and loneliness. Confession brings sin up into the open air, where it cannot breathe and will eventually suffocate and die. And so, confession is the way we ourselves come up for air, out of isolation and into true community. In our churches, we must teach, inspire, and equip the people we serve and lead to do this difficult yet necessary work, to learn and live out the art of confession. This work requires trust, which is developed over meaningful time spent together, often over a cup of coffee or a meal. And this means that we must create more spaces and opportunities to do just that.

STAYING AND FEASTING

There's a Korean dish I love called *moo guk*, which means "radish soup." It's a simple yet hearty soup made with radish, beef brisket, a lot of garlic, and an assortment of other ingredients. Eating it is always a nostalgic experience. It was a soup I ate not only at home growing up but also at church on an almost weekly basis.

I grew up in a Korean American church, and like most ethnic churches (and all Korean American churches), one of the primary gathering spaces for the community was around the dining table. I have fond memories of Sunday lunches after worship gatherings and of Friday dinners before youth group. Even as I type these words, I can see, hear, and smell in my mind that large fellowship hall, with a sea of people of all ages, amazing food, and the sound of chatter and laughter. These weren't quick and easy fifteen-minute meals, where you get in, get fed, and get out. Volunteers spent hours and hours preparing food, and people came and lingered for extended periods of time, breaking bread together. These weren't after-church meals. These meals *were* church.

In the digital age, we're losing our ability to slow down, stay a while, and feast. We do everything quickly and efficiently. Whether it's a restaurant or a church, we want to get in, get fed, and get out. This leaves little-to-no room for slowing down to develop meaningful connections with others, especially over coffee or a meal. But breaking bread together has been shown to connect us deeply in ways that no other experience can. A recent study from the University of Oxford showed that when we eat and drink together, our bonds are strengthened, we become increasingly content, and we experience a more intensified connection to the larger community.[15] Maybe this is one of the reasons why Jesus asked us to remember and celebrate him by gathering together over a meal.

Many churches I know in recent years have begun to place a heavier emphasis on this need for creating "stay and feast" spaces, which is a very encouraging sign. Within a thirty-mile radius of where I live, four different churches have opened coffee shops on their properties, open to the public seven days a week.[16] I've visited all of them several times and have been thrilled to see them full of people every time, Christians and non-Christians alike, gathering to slow down and enjoy coffee and conversation with one another. Many other churches I know in the area are beginning to lean into the food truck revolution by inviting local food trucks to camp out in their parking lots before and after worship gatherings.

I'm overjoyed to hear stories of these churches inviting people to share a meal together as a part of their worship gathering experience because I know firsthand what can happen. When we stay and feast together, strangers become friends, and friends become family. And while not every one of our churches can (nor should) open a coffee shop or invite local food trucks to take up parking-lot space, every one of our churches can take the basic principles and approaches of such spaces and incorporate key elements into our own spaces. As church leaders, it is imperative that we visit the places in our cities and towns where people seem to be gathering, be it a coffee shop, a food truck, a park, the town square, or someplace else, and ask the question, "What is it about this place that compels people to gather and stay a while?" As we ask the right questions, we'll find the right answers for our unique church communities.

PENTECOST ONLINE?

In Acts 2, we read about the birth of the early church: "When the day of Pentecost came, they were all together in one place" (Acts 2:1).

Let's stop there for a moment. We often gloss over this part of the story too quickly. At Pentecost, they were all together in one place. Read: *analog*. Some would argue, "Of course they were. They didn't have online platforms and video chats." Yes, true. But the fact of the matter is, they were together in one place and Luke, the storyteller, decides this detail is important enough to tell us so (even though it would've been assumed). It matters that they were all together in one place. Keep that in mind as the story continues:

> Suddenly a sound like the blowing of a violent wind came from heaven and filled the whole house where they were sitting. They saw what seemed to be tongues of fire that separated and came to rest on each of them. All of them were filled with the Holy Spirit and began to speak in other tongues as the Spirit enabled them. Now there were staying in Jerusalem God-fearing Jews from every nation under heaven. When they heard this sound, a crowd came together in bewilderment, because each one heard their own language being spoken. Utterly amazed, they asked: "Aren't all these who are

speaking Galileans? Then how is it that each of us hears them in our native language? Parthians, Medes and Elamites; residents of Mesopotamia, Judea and Cappadocia, Pontus and Asia, Phrygia and Pamphylia, Egypt and the parts of Libya near Cyrene; visitors from Rome (both Jews and converts to Judaism); Cretans and Arabs—we hear them declaring the wonders of God in our own tongues!" Amazed and perplexed, they asked one another, "What does this mean?" (Acts 2:2-12)

At Pentecost, as all the people are gathered together in one place, God imparts his Holy Spirit to his people, which enables them to speak in the native tongues of the many different people from all over the known world who had gathered.

Everyone is together in one place. God shows up. An unlikely group of people become one. This is how the church is launched. And every single piece of the story matters. What would've happened had there been a technological medium at the time to simulcast the event? Would the impact and influence of Pentecost have been the same for those who were simply "watching online" as it was for those in attendance? Unlikely. We see it in the way the people respond to the experience:

They devoted themselves to the apostles' teaching and to fellowship, to the breaking of bread and to prayer. Everyone was filled with awe at the many wonders and signs performed by the apostles. All the believers were together and had everything in common. They sold property and possessions to give to anyone who had need.

Every day they continued to meet together in the temple courts. They broke bread in their homes and ate together with glad and sincere hearts, praising God and enjoying the favor of all the people. And the Lord added to their number daily those who were being saved. (Acts 2:42-47)

Even after the fact, this unlikely group of people choose to stay together. They devote themselves to teaching and to fellowship. They feast together and pray together. They experience miraculous signs and wonders together. They share their resources with one another and work together for the common good. They provide for one another's needs through sacrifice and generosity. They continuously meet together. They eat in one another's homes and praise God together. And as they did all this, "the Lord added to their number daily those who were being saved" (Acts 2:47).

Of course. Who wouldn't be attracted to this sort of countercultural, subversive, selfless, generous, analog community? Nothing's changed. Our churches have the opportunity today to present this same countercultural, subversive, selfless, generous, analog vision of community to a world that longs for it. And as we do, I believe the Lord will again draw many who don't know him to himself through us.

A BROKEN-DOWN BOX TRUCK

When I was a youth pastor, every summer we would take our high school students to a lake for a week of wakeboarding, tubing, and swimming. One summer we combined with another youth group in town for the trip. We had over one hundred fifty students and leaders going to the lake together.

We also had a volunteer team coming to cook all our meals. We loaded an old, beat-up box truck with all our food and equipment for the week. Because the truck was a little shaky, I decided it'd be unfair to give the responsibility of driving it to a volunteer and decided I'd drive it myself. I brought along our then eighteen-year-old youth intern, Bryan. We set out about forty-five minutes before everyone else would leave, giving ourselves the extra time because of how slow this old box truck would be moving with so much weight in the back.

An hour into the drive, we were moving along and feeling good about things. Then all of a sudden, thick black smoke began to seep out of the engine. I pulled off the highway as soon as there was an exit and shut off the engine. Trying to exude a sense of calm in front of my intern, I said, "It's okay. Let's just give it a minute and then keep going." We were both rattled but hopeful it was just a simple overheating issue. After a few minutes, I started the engine and thankfully, we were back on the move. As Bryan and I laughed about what a close call that was, the engine started to smoke again. I pulled off at the next exit but this time, I didn't have to turn the engine off. The old box truck did that all on its own.

Now we were stuck. The facade of calm was gone. I began to visibly panic, which led Bryan to panic. Here we were, stuck on a highway, hours away from the lake, with all the food and equipment that one hundred and fifty people would need to survive for the week. I called a tow truck, which didn't arrive until nearly an hour later. While we waited, I looked out the passenger window and noticed a wide-open field with a distant horizon. I joked to Bryan that I wanted to just get out of the truck and run as far as I could down that field and never show my face anywhere ever again. Bryan looked at me and said, "Are you being serious? Because that's what I was thinking too."

The tow truck finally got us to a mechanic. A sense of relief set in. The truck was in good hands and we'd be back on the road in no time. Bryan and I walked to a diner across the street. Right as the onion rings were being set down, I got a call from the mechanic: "Hey, we don't have the parts the truck needs. So we're going to have to order them. It'll take a few days. Should be able to have the truck fixed in about a week or so."

A week or so? All the food and equipment were in that truck. We needed to get it up to the lake in a matter of hours, not days—certainly not a week. Then something strange happened. Bryan and I knew what needed to be done and we both felt an unexpected surge of energy welling up inside. We walked back across the street to the mechanic and I asked him to drive me to the nearest U-Haul. I told Bryan that while I was getting a U-Haul truck, I wanted him to unload as much of the box truck as he could. It's important to note here that this box truck had taken a dozen people about three hours to load. And now, I was asking a scrawny eighteen-year-old to unload it by himself. I went and rented a U-Haul and drove it back. To my surprise, Bryan had already unloaded about half of the box truck. I got out, jumped in the back and began helping him unload. Within thirty minutes, we'd unloaded everything. Another thirty minutes later, we'd loaded it all into the U-Haul. It wasn't pretty, and nothing was tied down.

Bryan and I were both bleeding and bruised. But the U-Haul was loaded, and we were back on the road.

The drive was smooth sailing from there. A few hours later we could see the lake off in the distance. It grew silent. It's hard to explain, but the emotions were overwhelming. Bryan began to cry. I wish I could stop there, but I can't, because when I saw Bryan crying, I began to cry. I don't know why. It's a little embarrassing, to tell you the truth. But you had to be there. You had to feel it and experience it and get your arms and legs cut up frantically unloading and reloading the trucks.

That's how the moments that most connect us to one another work. We show up, we go through the ups and downs, we face obstacles and difficulties, we get disappointed and beaten up and bruised, we bleed, we stick with it together, we overcome together, we accomplish the unthinkable together, we experience hope together, and then, finally, we arrive together. You have to be there.

PART 3

SCRIPTURE

SIX

JACKPOT!

SCRIPTURE IN THE DIGITAL AGE

In the choices we have made, consciously or not,

about how we use our computers, we have rejected

the intellectual tradition of solitary, single-minded

concentration, the ethic that the book bestowed on us.

We have cast our lot with the juggler.

NICHOLAS CARR

N COLLEGE, MOST WINTERS I'd make a snowboarding trip with friends up to the Lake Tahoe area. A few of my friends' parents were heavy gamblers and were regularly comped rooms, so we'd usually stay for free at one of the casino hotels. I wasn't a gambler (nor did I have any money to play with) but a few of my friends were. One night, my friend Edward spent an hour or so losing most of his money at a blackjack table and then, with a few nickels in hand, he walked over to a slot machine and started to play.

A few cents quickly turned into a few dollars. The few dollars eventually turned into twenty dollars. Then fifty dollars. Then a hundred. He was on a roll. A gleeful trance came over him as he fell into a strange sort of slot machine zone. Maybe the machine was broken, but it seemed like he couldn't lose. A few of us sat around watching him rack up jackpot after jackpot, until it got boring. We left Edward and went up to our room to watch a movie. Several hours later, around 3:00 a.m., we realized Edward hadn't come back. I went back down to the casino floor and found him sitting at the same slot machine, pulling the same lever, locked into the same trance-like zone. When I asked him how it was going, he told me that he'd won up to two thousand dollars at one point but was now back down to the original few nickels he'd started with. The jackpots stopped coming. The roll slowed to a standstill. The magic was gone. And yet there he was, still sitting at the slot machine, pulling the lever, asking me if I had any nickels.

Tristan Harris is the cofounder of an organization called the Center for Humane Technology, whose mission it is to "realign technology with humanity's best interests," born out of an awareness that technology often "tears apart our common reality and truth, constantly shreds our attention, or causes us to feel isolated."[1] Before his work with the Center, Harris worked as a design ethicist at Google. In 2016, he wrote an eye-opening article that laid out the calculated intentionality with which tech companies were designing applications to mimic the psychological effects of casino slot machines. Slot machines are built on a psychological dynamic called *intermittent variable rewards*, which work by rewarding a participant based predominantly on the number of times he or she executes a specific behavior. In other words, slot machines hypnotize us via the misguided belief that the more times we pull that lever, the more times we'll be rewarded. Be it once every three times, or once every thirty times, we keep on pulling because that's the only way to get closer to the next jackpot, regardless of when it may (or may not) come. According to Harris, this is exactly how many of the most popular digital applications work as well:

> When we pull our phone out of our pocket, we're playing a slot machine to see what notifications we got. When we pull to refresh our email, we're playing a slot machine to see what new email we got. When we swipe down our finger to scroll the Instagram feed, we're playing a slot machine to see what photo comes next. When we swipe faces left/right on dating apps like Tinder, we're playing a slot machine to see if we got a match. When we tap the # of red notifications, we're playing a slot machine to see what's underneath.[2]

It's no accident that the very physical act of refreshing many of the digital applications on our phones resembles the pull-and-release of slot machine levers. This is by intentional, methodical, well-researched design. These digital applications are meant to lull us into the same sort of trance-like zone I found my friend Edward in all those years ago in that dingy casino. In an article for the *Atlantic*, writer Alexis Madrigal explains the zone like this: "It's a rhythm. It's a response to a fine-tuned feedback loop. It's a powerful space-time distortion. You hit a button. Something happens. You hit it again. Something similar, but not exactly the same happens. Maybe you win, maybe you don't. Repeat. Repeat. Repeat. Repeat. Repeat. It's the pleasure of the repeat, the security of the loop."[3] Subtly and dangerously, our digital proclivities are transforming many of us into the sorts of people who pull the lever and refresh the page for no other concrete reason than to feel the pleasure of the repeat, to experience the security of the loop. Have you ever picked up your phone only to quickly and embarrassingly realize that you didn't actually have any specific reason to? So have I. We are all in some ways hypnotized by our digital slot machines.

Research is showing that the plasticity of the human brain is much greater than we'd previously known. In other words, our brains are constantly being reshaped by our experiences and behaviors. The more you pull the digital lever, the more your brain will desire another pull. And just like with casinos, the more you lose, the more you'll come back on the off chance that maybe next time you'll win. When we refresh our social media feed and discover that no one has liked or retweeted or shared our most recent post, we don't simply give up and never return. We refresh again, to see if maybe this time will be different than last time. And every now and then, it is.

This is the *intermittent variable rewards* dynamic at work. The problem is, the house always wins. And you will always lose. Pulling the digital lever may feel good in the moment by providing the occasional dopamine jolt you're looking for, but it will always lead, eventually and inevitably, to disappointment. And yet, in the digital age, we return to that grimy old chair in front of our digital slot machines, time and time again, without giving it a second thought. And in doing so, we are being changed, quite literally on a neurological level, into people who cannot stand to sink deeply and thoughtfully into anything that isn't as immediate as pulling the lever. Nicholas Carr describes it this way: "The Net's interactivity . . . turns us into lab rats constantly pressing levers to get tiny pellets of social or intellectual nourishment."[4] This is affecting us on a wide variety of levels but maybe none more alarming than the loss of the literary mind, the mind that reads deeply for extended periods of time to learn, experience, and grow.

A BOOKISH FAITH

A recent survey revealed that 25 percent of Americans admit to not having read a single book, in part or in whole, in the past year.[5] Not a single book read by at least one in four Americans. This is sad but not surprising. Thomas Pettitt, a professor at the University of Southern Denmark, writes that the age of reading books is "discernibly coming to an end under the pressure of developments in relation to the electronic media, the internet and digital technology."[6] This is a grim outlook. If we really are coming to the end of the age of reading books, if it truly was a mere phase in human history now being overtaken by the digital age, we're in very serious trouble. According to Carr, research is showing "that people who read linear text comprehend more, remember more, and learn more."[7] In returning again and again to our digital slot machines, we are losing our aptitude for reading, since reading requires "an ability to concentrate intently over a long period of time, to 'lose oneself' in the pages of a book, as we now say."[8] And in losing our aptitude for reading, we are also surrendering opportunities to comprehend, remember, and learn the deeper, more complex ideas and realities of human experience. But life is deep and complex, so to lose our ability to engage in these ways is to lose our ability to effectively pursue meaningful, purposeful lives. This is never truer than when it comes to Christian discipleship because the Christian faith has been and continues to be anchored by the sixty-six books of the Bible.

Scholar and historian Larry Hurtado calls Christianity a "bookish" faith, explaining that "both the importance and the impact of corporate reading of scripture writings are evident from the outset of the Jesus-movement."[9] The reading and corporate hearing of Scripture have always been vital parts of the life of Christian faith. The books of the Bible have always been at the center of the Jesus movement. Not only that, but up until just the last few hundred years, *reading the Bible* had primarily been a communal and extended act. It was communal in the sense that the corporate reading and hearing of Scripture was seen as the main way to engage the text. It was extended in the sense that these books were understood as long-format texts, meant to be read and heard either in their entirety or, at the very least, prolonged segments.

In the first century Greco-Roman world of Jesus, communal reading events were commonplace.[10] A very wide variety of written works were regularly read aloud in public gatherings—poetry, letters, historical narratives, speeches, wills. They were read in an equally wide variety of venues, often depending on the type of literature being read and heard—schools, dinner parties, theaters, courtrooms, and open fields. The Jewish people at the time of Jesus read and heard their Bible, the Old Testament, almost exclusively in synagogues, gathered with family, friends, and neighbors. The New Testament writers wrote most of their letters and stories with these sorts of audiences and environments in mind; these were texts designed to be read and heard out loud by a gathered people. And if people had gone to the trouble of gathering to receive a letter from the apostle Paul or Peter, would it make any sense to read only a handful of lines from the letter and declare, "Good for today. Let's finish it up tomorrow"? Of course not. The very nature of biblical engagement and the pragmatics of gathering in the early church led to focused and extended communal listening and receiving of Scripture.

By the sixteenth century, when the printing press placed Bibles in the hands of the masses, a once-scarce text only accessible by gathering was now made readily available. It gave individuals the opportunity to regularly access Scripture, conveniently and personally, for the very first time. No longer did individuals have to gather together at set times to read and hear the Word. They could do so when and where they chose, which paved the way for a more private engagement with the text. This newfound accessibility and convenience meant that individuals could bookmark their Bibles and come back to it when convenient. The need to read and hear as much as possible at one time was gone. You could now come and go as you pleased. The Bible would still be where you left it, waiting for you to pick up where you last left off. A few centuries later, the term *quiet time* was popularized by an InterVarsity booklet produced in 1945. Once Billy Graham began using the term during his evangelistic crusades in the 1950s, it quickly became one of the dominant ways Christians began to understand what effective engagement with the Bible looks like.

I do not mean to say that brief, personal devotionals have no value, and I am certainly grateful that Bibles are so easily accessible, at least in our part of the world. Quiet time alone with the Bible and your favorite cup of coffee in the morning, focusing on a short passage of Scripture can be a good thing. It's something I try to practice regularly. Most of the people of faith I admire most do the same, in their own ways. However, historically speaking, until recent years this approach has never been the primary mode of engaging the Bible.

Reading the Bible alone in short, bite-sized bits can be a healthy supplemental part of discipleship to Jesus but it must always be paired with an ongoing commitment to engaging Scripture as a whole, diving deeply into its long story, alongside the community of the church. But one of the greatest threats the digital age poses to Christian discipleship is the way the quick-fix lure of our digital slot machines is accelerating our penchant for engaging the Bible strictly as a fast, convenient, and individual exercise. That which was meant to be supplemental is quickly becoming our solitary, singular approach. While the digital age didn't start us on this trajectory, it does threaten to launch us even further in this direction at frightening speed.

Imagine a person who eats a well-balanced diet and takes supplemental vitamins to maintain health. Now imagine this person eliminates the well-balanced diet, which was at one point the primary means of her nourishment, and only takes the vitamins. No real food. Just vitamins. The results would be disastrous. When it comes to the way Christians read their Bibles in the digital age, replacing the primary with the supplementary approach leads to similarly disastrous results.

FILTERING THE BIBLE

As the digital age rewires us to crave the next push of the button or refresh of the page, we become increasingly incapable of engaging anything that requires deep, sustained engagement. As Carr writes, "What the Net seems to be doing is chipping away at my capacity for concentration and contemplation. Whether I'm online or not, my mind now expects to take in information the way the Net distributes it: in a swiftly moving stream of particles. Once I was a scuba diver in the sea of words. Now I zip along the surface like a guy on a Jet Ski."[11]

We who serve and lead in the local church must become increasingly aware and courageous enough to address the reality that the vast majority of the people in our communities are primarily jet skiing over the Bible rather than delving into its depths. They are very quickly and suddenly losing their ability to dive deep into the sea of not only words but stories and ideas. They see the Bible as just another available choice in the wide-open sea of their digital options and so, understandably, they approach it as they do everything else. Open the pages, pull the lever, see what comes up, refresh if necessary. The epic and expansive narrative of Scripture is rendered down to a series of disconnected morsels of encouragement or self-help suggestions. It's as if you took the text of a Shakespearean play, cut the lines into short single-sentence bits, and shoved them into individual fortune cookies. Sure, you'd be reading the same words, but you'd also be missing the story.

Examples of this downsizing of Scripture are common in the digital age. There are several popular Christian Instagram accounts, boasting hundreds of thousands of followers, that regularly post well-designed, aesthetically pleasing graphics of short, encouraging Bible verses. The complex story of Scripture is turned into a series of easy-on-the-eyes filtered images. Sometimes these posts may be what some people need to see at any particular moment, as long as they're seen as a supplementary source of encouragement and guidance. But far too often, these sorts of social media posts become the main source of spiritual nourishment for Christians in the digital age. Several of the more popular Instagram accounts have expanded to publish books and other Bible-reading resources. But even these expanded resources take a similar approach by attaching short, truncated biblical passages to inspirational devotional essays.

This gives the people who follow these ministries a false sense of spiritual nourishment without inviting them to feast on the story of the Bible as a whole.

Graphic artists will tell you that the ideas being communicated through visual media are about so much more than just the words. The way text is arranged and laid out says as much as the text itself. This is crucially important when considering the potential consequences of presenting Bible verses on social media. There are countless examples of how the design of a text miscommunicates or, at the very least, incompletely communicates its intended depth and breadth of meaning. One I saw recently was a quotation of Haggai 2:9, which says, "I will provide peace in this place." The design emphasizes the word "peace," in large font, centered against the backdrop of a perfectly blue and cloudless skyline behind a nondescript city building.[12]

The hope of God providing peace is wonderful and much needed in our world today, and it's also quite biblical. Many times, God promises and delivers peace in the midst of whatever chaos and pain we may be experiencing. But this verse is about so much more. In this somewhat obscure text, God speaks through the prophet Haggai and instructs his people to rebuild the temple, which in the Hebrew Scripture was the place where God would meet his people—the intersection of heaven and earth. This was during a time of exile, when the temple was in shambles. The verse referenced in the post is at the tail end of God's declaration that he would, "shake the heavens and the earth, the sea and the dry land . . . the nations . . . and fill [the temple] with his glory" (Haggai 2:6-7). The passage is about God's glory thundering down as the people rebuild the temple, culminating in the moment when God grants peace in that particular place, the intersection of heaven and earth where God would once again meet his people.

This may sound like Bible-thumping fundamentalism or overly analytical theological elitism to some but it's not. I'm simply pointing out the irrefutable contextual reality that God's promise to "provide peace in this place," at least in this case, is embedded in the story of his people rebuilding the temple. Does this mean that God doesn't promise to bring peace in other places and in other ways? Of course not. Does it mean that God doesn't long to bring peace into our personal lives? Of course he does. He has and will continue to do so. But it is important to understand the whole story of what's happening in Haggai, not simply because without it we misunderstand but more so because without it we miss out. Kevin Vanhoozer reminds us that "to pick and choose which words to heed and which to ignore is effectively to deprive those words of authority."[13] The authoritative power of Scripture is found not in individual verses but in the collective whole.

The comments on the post make clear that most people received the verse on a deeply personal level. There are "thank yous" for this wonderful reminder as people navigate relational conflict and weariness from the busyness of life. But the story in Haggai, when read and understood as a whole, displays a sort of peace that is much more expansive than one that simply brings calm to the conflicts and busyness of our individual lives. In fact, God's peace comes by way of the heavens and the earth, the land and sea, and all the nations shaken by God's glory. The peace of God covers so much more than just our personal lives. It certainly includes the details of what we're each going through, but it isn't exclusively for us. We experience God's peace most fully when it sweeps over the masses like a flood rather than arriving as a slow drip into our isolated circumstances. Additionally, in the story, the place where God's peace arrives is specifically in the temple, that fixed point where God meets his people and heaven collides with earth. But when taken at face value, a social media post like this is most often understood to mean that God's peace will be in whatever place I personally happen to be in that particular moment. Is that wrong? Not necessarily. But it certainly isn't what the story of Haggai is telling us.

When we splice Scripture this way and allow it to stand on its own, without context or an invitation to engage the entire story, we end up missing out on the learning, growth, and transformation that's only possible when we dive deep into the story from beginning to end, experiencing its ups, downs, and in-betweens. One of the great paradoxes of human experience is that we often become strong by experiencing weakness; we often become our best by experiencing the worst. As Cal Newport writes, "Human beings, it seems, are at their best when deeply immersed in something challenging."[14] One of the major problems caused by the way we're filtering the Bible in the digital age is that it teaches people to see the Bible only as a source of comfort. But the Bible is intended to both comfort and confront, in equal measure, as it should, for this is the way we grow. The filtering of Scripture almost always leaves out the confrontation as it picks and chooses to suit our personal preferences and needs. In turn, the process of discipleship is stunted. Taken to its extremes, the filtering of Scripture not only stunts our growth, it erodes it to critical levels.

READING FOR THE JACKPOT

I recently googled the phrase "What does the Bible say about" to see how its search algorithm—which is based on an aggregate of searches—would fill in the rest. This means that the top choices Google gave me are a basic summary of the most commonly searched topics by those looking for answers from the Bible. Here's what it looked like:

Q   what does the bible say about
Q   what does the bible say about **divorce**
Q   what does the bible say about **marriage**
Q   what does the bible say about **sex**
Q   what does the bible say about **cremation**
Q   what does the bible say about **tattoos**
Q   what does the bible say about **forgiveness**
Q   what does the bible say about **gambling**
Q   what does the bible say about **death**
Q   what does the bible say about **drinking**
Q   what does the bible say about **hell**

I did another search for the phrase "Does the Bible say" and here are those results:

> 🔍 does the bible say|
>
> ---
>
> 🔍 does the bible say **not to eat pork**
> 🔍 does the bible say **come as you are**
> 🔍 does the bible say **not to cuss**
> 🔍 does the bible say **not to eat shrimp**
> 🔍 does the bible say **not to drink**
> 🔍 does the bible say **anything about tattoos**
> 🔍 does the bible say **an eye for an eye**
> 🔍 does the bible say **angels have wings**
> 🔍 does the bible say **forgive and forget**
> 🔍 does the bible say **not to eat meat**

What this means is that when people go online to look for the Bible's answers, their most common questions are about things like divorce, marriage, sex, cremation, tattoos, forgiveness, gambling, death, drinking, hell, bacon, acceptance, cussing, and shrimp, among others. These results parallel my experience as a pastor. For the past decade and a half, almost every week I've been asked by one or several people in my church some form of the question, "Jay, what does the Bible say about . . . ?" I've personally been asked about every single result in those two Google searches; I've been asked about most of them many times.

Going to the Bible for answers is okay, but it's not what the Bible is primarily for. The Bible is not an IKEA manual with step-by-step instructions on how to build a wonderfully efficient life. It is a library of sixty-six ancient books, written in several languages, all of them foreign to most of us, in times and cultures most of us know very little to nothing about. Any answers we may find in Scripture will not come simply or easily. Furthermore, the biblical narrative is not so much in-terested in giving us pat answers as it is in helping us ask better questions. In other words, the Bible is not a slot machine with a lever we pull at our convenience, hoping to hit the answer jackpot. It's far too alive and God-breathed (2 Timothy 3:16), to be that mechanical. It's far more confrontational and provoc-ative than the self-help guide we want it to be. And yet, this is how so many people approach Scripture today.

This is in large part due to the profound and dramatic influence of the digital age on the way we think about and engage not only the Bible but also our search for knowledge and information on the whole. Nicholas Carr writes, "We are evolving from being cultivators of personal knowledge to being hunters and gatherers in the electronic data forest."[15] This is exactly how so many Christians today are engaging Scripture—as hunters and gatherers, quickly and efficiently flipping through the pages, on the prowl for whatever prey they happen to need in the moment, be it easy answers to life's hard questions, encouragement, in-spiration, or self-help tidbits. But this sort of calculated ap-proach to the Bible leads us down the path to missing the point entirely.

Alan Jacobs says that good books ask us to "change our lives by putting aside what we usually think of as good reasons. It's asking us to stop calculating."[16] The Bible, of course, is more than just a good book. It's a library of the best and most important books in human history. More than that, it's God's holy Word. So we must stop calculating and start reading—deeply, patiently, together. We must stop reading for the jackpot and begin the work of diving in deeper.

LEARNING TO SWIM IN THE DEEP END

In the little apartment complex where my mother and I lived for a few years when I was growing up, we had a swimming pool. The pool was outside our back door, which was convenient for me in the summers. I'd swim for hours every day. No one ever taught me to actually swim, so when I say I swam for hours, what I mean is, I splashed around idly in the shallow end by myself. Sort of sad, now that I think about it.

Then one day, I was invited to a friend's birthday party at a local water park. Thinking I was an experienced swimmer from all my time in the shallow end, I went to the party, confident and excited. The first part of the park my friends and I went to was one of those giant pools with an island in the middle. There were ropes hanging above at the edge of the pool, which you'd grab onto to make your way to the island. Without hesitation, I jumped up, grabbed the rope, and began to make my way to the island. But I soon lost grip and found myself in the water. The deep water. It was then that I realized, *Oh, I don't actually know how to swim.*

The feeling of drowning is frightening. I remember wondering two things: *Is my mom going to have to eat dinner alone now? And what will she do with all of my G. I. Joes?* Just as I thought this would be the end, a young lifeguard lifted me up out of the water and gave me a new lease on life. A couple of days later, back at home, I decided that I needed to learn how to actually swim. I began by jumping into the shallow end of the pool but not allowing my feet to touch the floor and seeing if I could make my way back above water. I did this for several days, until I got comfortable. Then one day, I stood at the edge of the deep end. I took a breath, hesitated, took another deep breath, then jumped in. To my surprise and joy, I swam my way back above water.

My friend Tim Mackie of the Bible Project describes the Bible as "a unified story that leads to Jesus."[17] To read the Bible as anything less than that is like swimming in the shallow end of the pool. We can splash around and feel just fine for a while. But at some point we're going to be thrown into the deep end. Whether it's our own life circumstances or those of loved ones, the deep end is coming if it has not yet already come, both for you and for the people you serve and lead. And if we are forever swimming at the shallow end of the pool, we will forever be shallow Christians, ill-equipped to handle the doubts, fears, and questions that will surely find us. And discipleship to Jesus itself is a deep work, as we've established in previous chapters, not a shallow one. Not only that, but we also live in a post-Christian world in which the dominant culture presupposes that the Bible is an archaic, outdated, even inhumane work of ancient literature that ought to be ignored at best and banned at worst. Swimming in the shallow end of the Bible not only ill equips people to respond effectively but, more insidiously, it exposes them to completely drowning in the deep end of culture's overwhelming questions and critiques. People with brilliant minds such as Richard Dawkins, one of the world's leading atheists, are positing ideas like this one: "The God of the Old Testament is arguably the most unpleasant character in all fiction: jealous and proud of it; a petty, unjust, unforgiving control-freak; a vindictive, blood-thirsty ethnic cleanser; a misogynistic, homophobic, racist, in-fanticidal, genocidal, filicidal, pestilential, megalomaniacal, sadomasochistic, capriciously malevolent bully."[18]

This is the deep end. Ironically, Dawkins comes to this con-clusion because he's reading the Bible exactly the way so many Christians are—in bite-sized morsels, detached from the epic and beautiful scope of the entire narrative arch. He's simply pulling the lever and coming up with different types of jackpots than the Christians who read in the shallow end themselves. But the approach is the same.

So what must we do to swim in the deep end of Scripture for ourselves and invite, encourage, and equip the church communities we serve to do the same? If reading the Bible in the digital age has become about the lure of the digital slot machine, pulling the quick-fix lever to get what we want, then what does it mean to engage the Bible in analog?

We must become slow. That's what an analog approach to the Bible is all about. So please, keep reading . . . this book, and the Bible.

SEVEN

HOWTOREADABOOK

ANALOG SCRIPTURE

The more that you read, the more things you will know.

The more that you learn, the more places you'll go.

DR. SEUSS

S EVERAL SUMMERS AGO I spent a couple of weeks in Kenya, visiting churches and helping support some of their local ministries. One Sunday we attended a worship gathering at a church we were partnering with, and afterward they invited us over to one of the families' homes for lunch. When we arrived, we learned that several families in the church had gotten together to cook an elaborate meal for us. The main course was a goat stew that blew my mind. It was delicious in a way that was new and hard to describe but undeniably one of the best things I've ever tasted. As we were chatting over lunch, I asked the older woman who was responsible for the stew about her recipe, if she had any secret ingredients. She smiled warmly and said to me, "Slow." Confused, I asked what she meant. She said it again: "The secret ingredient is 'slow.'"

In addition to the various spices and seasonings, the meat itself, and even her skilled cooking, what really brought the stew together was her willingness to take it slow. This is true of any great dish. The best meals are always prepared patiently and carefully over many hours because there's just no adequate substitute for slow cooking. The only way to draw out the full depth and richness of flavors in most dishes is to give them time.

The same holds true when it comes to the Bible. The only way to experience the full depth and complexity of its unfolding story is to give it time. In the digital age, our tendency is to microwave everything. We've grown impatient and we have a hard time waiting. But in order to experience the Bible at a deep level and allow it to do its work in us, we must understand that it will take dedication, devotion, and commitment over the long haul.

The secret ingredient is "slow."

LEARNING TO READ SLOWLY

A little while back, a friend of mine paid a lot of money to take a speed-reading class. I asked him what his motivation was, and he told me succinctly, "So I can read more in less time." The math is simple enough. Shortly after, I read an article in *Wired* magazine written by Mark Seidenberg in which he critiques the entire concept of speed reading.

"There is one simple, guaranteed way to increase reading speed: skimming," he writes. "There is a trivial sense in which these texts are being read rapidly, but very little is being comprehended. We should call this Quote-Unquote Reading or Sorta Reading rather than speed reading."[1] Seidenberg calls it the "myth of speed reading" because, really, there is no such thing.

This isn't conjecture on his part. It's science. The interactive partnership between the eyes and the brain can increase its capacity for seeing words at greater speeds, but it is limited in its capacity for comprehending the meaning of these words. Seidenberg writes, "The exact number of words per minute is far less important than the fact that this value cannot be greatly increased without seriously compromising comprehension."[2]

The reading platforms of the digital age, exclusively presented to us on screens rather than pages, with neatly uniform rows of text and just the right amount of spacing, are, by design, meant to help us read simply and swiftly. Speed is one of the key values of the digital age, so it's no wonder that its written text is designed to be read speedily. This is pragmatically helpful, but there is a loss: namely, the loss of deep comprehension.

Ancient texts were written very differently. There was no thought given to the pragmatism of reading simply, swiftly, or speedily. And this wasn't seen as a problem or an issue. In his book on the history of scribal text called *Space Between Words*, Paul Saenger writes, "Because those who read relished the mellifluous metrical and accentual patterns of pronounced text, the absence of interword space in Greek and Latin was not perceived to be an impediment to effective reading, as it would be to the modern reader, who strives to read swiftly."[3] This is emphatically true of the original manuscripts of the Bible. What we know as the Old Testament was originally written in Hebrew, with a few short sections in Aramaic.[4] In both the original Hebrew and Aramaic texts, there are no vowels or punctuation. While there is some spacing between words, it is often minimal and not easily decipherable. And the earliest New Testament manuscripts were written in Greek, in all caps, with no spacing between words and minimal punctuation.

THINKABOUTREADINGLIKETHAT

IMAGINEHOWSLOWLYANDCAREFULLYYOUWOULDHAVE TOREAD

SERIOUSLYTHINKABOUTIT

How long did it take you to read those last three lines? Longer than normal, I assume. As you read, whether you knew it or not, something was happening neurologically. As your eyes slowed down to decipher the text, so did your brain. Each word was given an extended amount of time to linger and bounce around in your mind. Once you realized the full structure of each sentence, your mind had spent exponentially more time processing the words than you would reading normal modern text. As Nicholas Carr notes, reading like this is like "working out a puzzle. The brain's entire cortex, including the forward areas associated with problem solving and decision making, would have been buzzing with neural activity."[5]

This is how people read for the majority of literate history. Reading was an act of problem solving and decision-making. It involved a strenuous flexing of the mind's muscles and required deliberate and focused concentration.

But in the digital age, a growing majority of our reading is being done in abbreviated forms. On one hand, this is a tremendous advantage, in that we can so efficiently access interesting information in a few brief two hundred and forty-character blurbs. Yet it is imperative that we recognize the limitations of this type of reading. While they are sometimes adequate for information, social media threads and brief blog posts often lack the necessary breadth to navigate well through complexity and nuance. But deep rooted change via the written word only happens when it takes us through complex and nuanced waters; therefore, we must learn to read slowly if we are going to experience any sort of transformation through a particular text. As I wrote in the previous chapter, the problem is that the digital age is changing our aptitude and appetite for this type of deliberate and concentrated reading. The more we read in abbreviated forms, the more we long for that sort of reading and the less able we are to read slowly.

Think about how odd it would be to receive a text from someone with the phrase "Laughing out loud!" Wouldn't your first thought be, "Why would you write all of that? Don't you know 'LOL'?" That's the digital age. Now, here's the rub: when we read "LOL" in our text messages, we are acutely aware that there is a high probability the person on the other end isn't actually laughing out loud. In fact, people will often text one another "I'm literally LOL-ing!" as an admission that normally "LOL" isn't literal.

For the entire history of the English language, the phrase "I'm laughing out loud" had a singular concrete meaning. But the digital age has changed that. First, it's made the phrase itself almost obsolete. Second, it's deconstructed the phrase, detached it from its original definition, co-opted it into the abbreviated "LOL," and infused it with a much broader spectrum of meaning. The point here is that the digital age's effect on reading is in turn having an effect on language and meaning, which in its turn has a profound effect on our intellectual lives. Carr writes, "Because language is, for human beings, the primary vessel of conscious thought, particularly higher forms of thought, the technologies that restructure language tend to exert the strongest influence over our intellectual lives. . . . The history of language is the history of the mind."[6] How we read affects how we communicate, and how we communicate affects how we think, and how we think affects who we become.

So what does all of this mean for us and our relationship with the Bible? It means that we must unlearn the skill of speed reading and relearn the basics of how to read a book, slowly.

HOW TO READ A BOOK

In 1940, author and philosopher Mortimer Adler published his famed *How to Read a Book*. In 1972, he cowrote and published the second edition of the book with writer Charles Van Doren (who was portrayed by Ralph Fiennes in the underrated 1994 film *Quiz Show*, but that's neither here nor there).

The book is considered a classic and is often referenced by academics and general audiences alike as a seminal work on the topic. In the book, Adler and Van Doren present four key questions that all good readers must ask of any book:

1. What is the book about as a whole?

"You must try to discover the leading theme of the book, and how the author develops this theme in an orderly way by subdividing it into its essential subordinate themes or topics."[7]

2. What is being said in detail, and how?

"You must try to discover the main ideas, assertions, and arguments that constitute the author's particular message."[8]

3. Is the book true, in whole or part?

"You cannot answer this question until you have answered the first two. You have to know what is being said before you can decide whether it is true or not. When you understand a book, however, you are obligated, if you are reading seriously, to make up your own mind."[9]

4. What of it?

"Why does the author think it is important to know these things? Is it important to you to know them? And if the book has not only informed you, but also enlightened you, it is necessary to seek further enlightenment by asking what else follows, what is further implied or suggested."[10]

These are the right questions to ask. The order is of the utmost importance, but so often, in our approach to Scripture in the digital age, the questions have been reversed. People reading the Bible usually begin by asking, "What of it?," the fourth and last question. In other words, they approach the Bible seeking enlightenment, searching for clear implications or suggestions, without having done the prerequisite work. This is the "reading for the jackpot" approach: seeking self-help tidbits or small morsels of encouragement or inspiration for the day. But the danger of this reordering of the questions is obvious. How could we possibly embrace the big ideas, implications, and suggestions of the Bible without knowing if they're true? And how can we know if they're true without first doing the work of discovering the details of the story— the main ideas, assertions, and arguments, as Adler and Van Doren point out? And how can we accurately understand the details without first understanding the overall theme of the book as a whole, the backdrop which gives context to those details? In the digital age, one of the most important things we can and must do in helping the people in our communities more deeply and dynamically engage the Bible is to teach them to read slowly enough to thoroughly ask each one of these questions, in the right order.

WHAT IS THE BOOK ABOUT AS A WHOLE? THE BIBLE AS A UNIFIED STORY

The broadcast era gave rise to a shift in the way churches approached sermons. Following the example of broadcast television programs, sermons began to take on formulaic structures—an opening, a text, some exegesis, an anecdote or personal story, an emotive invitation, and so on. Church leaders began to see the Bible as source material for weekly thirty-minute episodes. This shift eventually led to what some call the "felt-need" sermon series. The formula is simple. Identify a felt need in people's lives. Map out a sermon series addressing the key topics within that felt need, week by week. Find suitable, bite-sized Bible passages to fit into the sermons.

While I am not against felt-need sermon series (we usually have a few of them a year at our church), I believe the approach must change. I am not suggesting that churches should all move to a lectionary-driven model of preaching (although that can be a great approach in many contexts). I think there is still a place for felt-need–type sermon series in many if not most of our church contexts. What I am suggesting is that in approaching felt-need sermons, we must maintain and uphold faithfulness to the overarching narrative of the Bible. We must never forget and constantly remind our people that the Bible is a unified story, where the beginning, middle, end, and everything in between are all connected. We must teach and preach Scripture in a way that points people to the pinnacle of the story, the life, death, and resurrection of Jesus Christ, always. We must be mindful of how easy it can be to slip into the "reading for the jackpot" approach, where the Bible becomes a series of individual fortune cookies rather than a single compelling story. We must remember, as Dietrich Bonhoeffer reminds us, "The Holy Scriptures are more than selected Bible passages.... The Holy Scriptures do not consist of individual sayings, but are a whole and can be used most effectively as such. The Scriptures are God's revealed Word as a whole."[11]

This can be done a number of ways. First, regularly reading, teaching, and preaching through entire books of the Bible in our communities must always be a part of the rhythm of church life. This requires us to prayerfully and carefully map out what the year or years ahead will look like in terms of preaching and teaching. Second, when we do craft felt-need sermon series, they should be born out of and embedded within a portion of the biblical narrative. One church I know recently taught a felt-need series on marriage, but rather than picking and choosing a variety of passages, they taught through entire sections of the book of Genesis. Another church I know, when teaching a felt-need series on evangelism and sharing one's faith, anchored the entire series on the book of James. This sort of approach requires intentionality, creativity, and flexibility, not to mention a prayerful seeking of God's wisdom and insight. But it's possible. We must never forget that the biblical story is at its core a story of human experience, what it was meant to be, how things went awry, and what God is doing to make things right again. This means that there is not a single human felt need that the Bible can't breathe hope and life into. But we can't take shortcuts getting there. As Scot McKnight says, "Just as shortcuts in exercise prevent full health benefits, so also shortcuts in Bible reading affect our spiritual health."[12] We must do the work.

Helping the communities we serve approach Scripture as a unified story goes beyond the sermon as well. One of the most important things we must do is to encourage people to actually read the Bible from beginning to end. This can be daunting, of course, so it's important to equip them well. There is an assortment of resources out there, but I can't think of a better one than The Bible Project.[13] I quoted my friend Tim Mackie in the previous chapter. Tim cofounded the Bible Project along with Jon Collins several years ago and today it is arguably the most influential online Bible resource in the world. They create incredibly dynamic and compelling short-form animated videos that explore various books and themes of the Bible, as well as a wide variety of theological ideas. All their videos are available for free at their website and on their YouTube page.[14] They also have a Bible-reading plan on their website, which includes suggested videos to support the reading.

The Bible Project also did something remarkable in December of 2016. As an experiment, they decided to host a book of Revelation reading event. The event was exactly what it sounds like—a straight read-through of Revelation in its entirety. As a part of the night, they premiered their video on Revelation. Tim tells me they had no idea how many people would show up. But on a cold December evening in Portland, more than four hundred people came to hear the book of Revelation.[15] No music. No sermon. No Christian-celebrity guests. Just ninety minutes of the last book of the Bible, beginning to end, the way the original audience would've heard it.

Inspired by this, several churches I know have begun Read Scripture events. Some are one-off events where people gather to read and listen to an entire book of the Bible read aloud. Others, like the one at Neighborhood Church in Visalia, California, are weekly gatherings. At Neighborhood, every Friday morning at 6:00 a.m. dozens of people gather to read through the Bible together. Nothing is provided other than coffee, a warm room, and the Bible. And they come in droves.

All of this seems counterintuitive, but in the digital age, people are hungry for analog experiences like this. Especially when it comes to the Bible, one of the most counter-cultural—and therefore intriguing and attractive—things we can do is invite people to slow down, settle in, and engage the whole unified story from start to finish, no matter how long it takes.

WHAT IS BEING SAID IN DETAIL, AND HOW? AND IS IT TRUE? READING THEOLOGICALLY

As people begin to read Scripture as a unified story, questions will inevitably arise. The Bible is a library of ancient books written, transcribed, compiled, and translated over the course of several millennia, so to think that it wouldn't lead to a plethora of questions and confusions would be naive.

This is where theology comes in. Our tendency today is to relinquish theological responsibility and put the onus on paid Christian professionals to do the work for us. But in reality, we are all theologians. As A. W. Tozer famously wrote, "What comes into our minds when we think about God is the most important thing about us."[16] This is true for all people—atheists, agnostics, religious types, and everyone in between—because whatever we do or don't believe about God—that he doesn't exist, that he can't be known, that he's my personal Lord and Savior, or whatever else—is the foundation upon which the rest of our worldview is built. This means that we are all doing theology all the time. Everything we think, say, and do is theological because it's all built on the foundation of what we think we know or don't know about God.

This work of theology is intricately linked with biblical exegesis.[17] For our purposes here, we will borrow again from Adler and Van Doren in describing biblical exegesis simply as the process of diving into the details of what's being said and how it's being said. The important thing is to remember that, just as is the case with theology, "everyone is an exegete of sorts. The only real question is whether you will be a good one."[18] When we encourage people to read the Bible, we are also encouraging them, by default, to exegete. If we do not equip them to do it well, they will exegete anyway, often without even knowing it, by inferring meaning and ideas into the text based on their own experiences, intuitions, personalities, and preferences. So again, we must create spaces and opportunities for them to slow down and settle in to responsibly engage the Bible theologically, through exegesis. There are many ways to do this, but I'll suggest a couple of ideas and resources.

Several churches in the past few years have embarked on a year-long journey through the Bible and theology called the Year of Biblical Literacy.[19] Born out of a partnership between Reality San Francisco, Bridgetown Church, and the Bible Project, the Year of Biblical Literacy is a calendar year scheduled and mapped out to help churches engage key biblical and theological themes together. Their website provides free resources for pastors and church leaders, small groups, and individuals. The resources include sermons and lectures, small group discussion guides, reading plans, and experiential practices for groups and individuals. All of the several churches that I've spoken with who have gone through the Year of Biblical Literacy say that it's had an incredibly positive impact in their communities.

In 2013, my friend Dan Kimball partnered with Western Seminary to create the ReGeneration Project.[20] Motivated by the rapid decline in biblical literacy and theological engagement, particularly among younger generations, the project was launched to help young people and the church leaders who serve them tackle the confusing and complicated stuff of the Bible together. For the past several years, I've had the privilege of serving on the leadership team, where I help to develop training events and cohost a theology podcast with my friend Isaac Serrano, another local church leader who is passionate about reaching new generations. We believe that theology is for everyone, so our website and events offer accessible, understandable theological and biblical content to help the nonacademic, non-seminarian engage deeply.

Whether it's the Year of Biblical Literacy, the ReGeneration Project, or something else altogether, what's most important is that in serving and leading our church communities, we must carve out regular space and create regular opportunities for training and equipping when it comes to the Bible. We have to begin creating a culture of learning. As I wrote in the previous chapter, if we are not swimming in the deep end of the pool, the deep end will still someday come to us. It is our responsibility to equip people to not only keep their heads above water but to swim well. The Bible is challenging in many ways, and to deny the questions, doubts, and anxieties it can raise is to deny its beauty, goodness, and power, as they are inextricably tied to one another. Oliver Wendell Holmes is quoted as having said, "I would not give a fig for the simplicity on this side of complexity, but I would give my life for the simplicity on the other side of complexity." To arrive at simple answers without first wading through complex waters is to find no answers at all. So we must invite our people into the slow and steady waters of theology and exegesis. We must meet them there and journey with them toward rich and robust simplicity on the other side. This is difficult work that cannot be done alone, which is one of the reasons why the church matters so much. Especially in the digital age, our churches must become learning communities.

WHAT OF IT? READING TRANSFORMATIONALLY

The final question Adler and Van Doren ask is in many ways the most important one: "What of it?" What's the point? Beyond knowledge and understanding, what is our engagement with the Bible supposed to actually do? What does it affect or change?

The answer is, simply put, *everything*. The biblical word for this is *transformation*, as in Paul's letter to the Romans: "Do not conform to the pattern of this world, but be transformed by the renewing of your mind" (Romans 12:2). In the original New Testament Greek the word here is *metamorphoō*, which suggests a deep change in the actual form or nature of a person or a thing. Think a caterpillar emerging from its cocoon as a butterfly. This is transformation. "The old has gone, the new is here!" (2 Corinthians 5:17). This is the declarative answer to the question, "What of it?" Ultimately, we read the Bible to experience transformation, to be changed completely, from an old thing to a new thing. Anything short of that would be to miss the point.

We can know and understand the unified story of the Bible like the back of our hands. We can become the sorts of people who achieve doctorate degrees in theology and master the art of exegesis. But if our engagement with the Bible doesn't transform us by drawing us closer to God and remaking us into the image of Christ, then it's all for naught. The missiologist Lesslie Newbigin once wrote, "The Bible is the way in which we come to know God, because we don't know a person except by knowing his or her story."[21] In this way, Scripture is the journey, but a life transformed into more Christlikeness is the destination. If we're not careful, our pursuit of knowledge and understanding can easily lead us down the path of misguided intellectual ascent. If we are unaware, we can fall into the trap of theological elitism, where having the right biblical answer becomes the goal, rather than becoming the right person in light of who Christ is, what he's done, and what he's calling us to become. We can easily find ourselves in hot pursuit of Bible mastery.

But as Scot McKnight reminds us, "God did not give the Bible so we could master him or it; God gave the Bible so we could live it, so we could be mastered by it. The moment we think we've mastered it, we have failed to be readers of the Bible."[22] Reading the Bible well results in being mastered by it, not the other way around. That's the incredible paradox of Scripture, which exists because the books of the Bible aren't just any ordinary books. Unlike the wonderful yet inanimate objects filling our libraries and bookstores, the books of the Bible are alive in ways that no other books in history are or ever will be. This is both a claim and a promise that the text itself makes: "For the word of God is alive and active. Sharper than any double-edged sword, it penetrates even to dividing soul and spirit, joints and marrow; it judges the thoughts and attitudes of the heart" (Hebrews 4:12) and "All Scripture is God-breathed and is useful for teaching, rebuking, correcting and training in righteousness, so that the servant of God may be thoroughly equipped for every good work" (2 Timothy 3:16-17).

When we encourage and equip people to dive into the deep end of Scripture, we are inviting them into a living, breathing experience of transformation. We are not asking them to merely read a book or study a text. We are asking them to enter a relationship. We're asking them to be vulnerable, honest, teachable, and willing. And if we aren't asking these things of them, then we're failing to help them read transformationally, which is exactly the way the Bible is intended to be read; and not only read, but received, embodied, and lived out. When people engage the Bible most fully, it will result in not only an expansion of knowledge but, more importantly, a reorientation of the will, from self-centric motivations to a posture of obedience and surrender. A transformed will is one of the surefire signs that Scripture is doing deep work in someone's life. Dallas Willard wrote that, "The will to obey is the engine that pulls the train of spirituality in Christ."[23]

This is the tremendous task at hand for those who serve and lead in the church. We are called to first and foremost engage the Bible for ourselves in this way, allowing it to do deep work in us, transforming us day by day. Coupled with that, we are called to embody and express that journey of transformation through Scripture in such a way that those in our church communities might find, in witnessing our experiences, the necessary courage, patience, and trust to read transformationally for themselves.

PASSING IT ON

As far back as I can remember, on the nights that my mother was home (she often worked two or three jobs, so those nights weren't as often as she would've liked), we would read the Bible together. She called it "family worship time." Each night before bed, she'd sit me down at the kitchen table and we'd go through the same routine. I'd say a trite prayer, usually no more than a couple of sentences. Then we'd open our Bibles and read extended sections out loud. Sometimes it was a chapter, occasionally it was several. Then she'd pray, never trite like me, always passionate and specific. Finally, we'd end by reciting the Lord's Prayer together.

I have to tell you, for most of my childhood I absolutely despised those last fifteen-to-twenty minutes of the day. Many a night, I lay in bed early with my eyes closed, in hopes that she'd assume I'd fallen asleep and leave me alone. But she always knew. And every night, I'd drag myself to the kitchen, sit myself down, and Eeyore my way through yet another family worship time.

In hindsight, I have a much better understanding of why my mother did what she did, and I am forever grateful. I'm grateful because she took seriously the biblical call to pass on the story of God to the next generation. She happily shouldered the responsibility of sacred storytelling, doing everything she could to make sure Scripture would saturate my heart and mind. To this day, my mother begins and ends her day with the Bible and prayer. The magnitude of her faithfulness to God is inexpressible, and the seriousness with which she takes Scripture is inspiring. So it's no wonder that even in the midst of the busyness and exhaustion of life as a single mom working multiple jobs, she would still carve out time every night to read the Bible with me. It's no wonder because she took so seriously passages like Joel 1:3, "Tell it to your children, and let your children tell it to their children, and their children to the next generation," and Psalm 145:4, "One generation commends your works to another; they tell of your mighty acts."

Sadly, in the digital age, we seem to be fumbling away the responsibility and privilege of passing on the story of God to younger generations. This is doubly disconcerting because the trends seem to indicate that younger generations today are the most biblically illiterate in American history.[24] While older generations often cast blame on younger generations for not caring enough or being too dismissive, the reality is that in the digital age, we are all collectively losing sight of the power of sacred storytelling and the crucially important need to pass on and receive the story of God from generation to generation. This too is in many ways a direct result of the digital age. In her book *Reclaiming Conversation*, Sherry Turkle writes,

Technology enchants; it makes us forget what we know about life. The new—any old new—becomes confused with progress. But in our eagerness, we forget our responsibility to the new, the generations that follow us. It is for us to pass on the most precious thing we know how to do: talking to the next generation about our experiences, our history.[25]

Passing the faith on to the next generation isn't easy or convenient. It requires sacrifice and focus and compassion and patience. It is often a slow work. But as we've seen, the Bible, much like a great stew, is best experienced and engaged slowly. The depth, richness, and complexities of our faith are only revealed adequately over time. This is to our advantage as we think about passing on this faith to the next generation.

These days, I have a toddler of my own. Most nights we sit on her bed, pray, and (try to) read the Bible together. Most nights, she's distracted and eager to be done. It's a nuisance to her. She has other things she wants to do, games she wants to play, books she wants to read, a baby brother she wants to bother. I know what she's going through. But I also know something she doesn't. I know that this slow process, one that feels excruciatingly so to her fast-twitch four-year-old mind, is exactly how the Bible is meant to be experienced. And I am hopeful that, in the end, if we're faithfully consistent and show up night after night, in the words of the psalmist in Psalm 34:8, she too may one day finally, "taste and see that the LORD is good." Sure, it may take a while. And of course it will. The secret ingredient is "slow," after all.

EIGHT

THE MEAL AT THE CENTER OF HISTORY

COMMUNION

We need more than talk, more than words on

a page; we need a touch, a smell, a taste. . . .

The sacrament is more than a medium of communication;

it is a medium of action, God's action.

LEONARD VANDER ZEE

RECENTLY LOST MY WEDDING RING. I searched everywhere, high and low, but it was nowhere to be found. My wife and I took sheets off beds and cushions off couches, and we turned over pots and pans. We searched for two days and couldn't find it anywhere. I felt strangely off without it. My finger felt weightless—empty and exposed in a weird way.

Then one morning, my then three-year-old daughter grabbed my hand, saying, "Daddy's ring, Daddy's ring," and walked me to her playroom. There, she pointed down to the bottom of one of her toy bins. There it was. My ring. She'd taken it off my nightstand a couple of days earlier while I was washing up and dropped it in with her unending pile of toys. I felt an unexpected sense of wholeness come over me. But why? It's just a ring, right?

In October of 2008, the great American theologian Beyoncé released one of her most important works, titled "Single Ladies," in which she reminded us, "If you like it, then you shoulda put a ring on it." I took these words to heart, and just three short months later, I stood across from Jenny in front of our closest family and friends and did exactly that. I put a ring on it. And she put a ring on me. And those rings, for both of us, are worth more than the dollars and cents they cost. Their value isn't tied to their material worth but to something much deeper and more significant. While most would call these rings mere symbols, they seem more than that to me. Their physicality, the binding up of skin and bone, the added weight to our fingers are all important, not only as a form of reminder but also as a constant, relentless reality, worn on our bodies.

In the digital age, the question most are asking is, "What's the maximum amount of efficiency and convenience we can achieve?" In other words, in the digital age, what's most efficient and convenient is always preferable. And so, technology offers us an endless assortment of online and virtual substitutes for real things and experiences because they're deemed more efficient and convenient. But sometimes efficiency and convenience aren't actually best. As physical, sensory creatures, human beings need physical, sensory reminders and realities. Thoughts, feelings, beliefs, and ideas sometimes need to be, quite literally, fleshed out. This is especially true when it comes to the most important things in life.

For Christians, the most important thing is the new life we have and live in Jesus. As such, it is not enough to carry this new life around with us as a thought, feeling, belief, or idea. We also need something we can see, smell, taste, and touch, to both remind us of this new life and to bring it to bear as a present reality. Leonard Vander Zee writes that "we need physical and material confirmation of our new relationship with God in Jesus Christ because we are physical and material beings."[1] This is why, on the night before his arrest and eventual execution, in summarizing the work and intent of his earthly ministry, Jesus did not choose to merely verbalize his thoughts, feelings, beliefs, and ideas. Instead, he shared a meal with his closest friends: "While they were eating, Jesus took bread, and when he had given thanks, he broke it and gave it to his disciples, saying, 'Take and eat; this is my body.' Then he took a cup, and when he had given thanks, he gave it to them, saying, 'Drink from it, all of you. This is my blood of the covenant, which is poured out for many for the forgiveness of sins'" (Matthew 26:26-28).

Eating and drinking. Two of the most foundationally analog practices in human experience. This is how Jesus culminates and summarizes his life and work. Together, Jesus and the disciples eat and drink, they smell, touch, and taste their way to the new reality that is about to come. And this is exactly what so many followers of Jesus have lost and thus more desperately need in the digital age.

EATING AND DRINKING OUR WAY INTO COMMUNITY

This final meal before Jesus' crucifixion is known by many names today—Communion, Eucharist, the Last Supper, Mass.[2] It's important to remember, however, that this meal was first and foremost a traditional Jewish meal known as Seder, the ritual feast that marks the beginning of the Passover festival: "When the hour came, Jesus and his apostles reclined at the table. And he said to them, 'I have eagerly desired to eat this Passover with you before I suffer. For I tell you, I will not eat it again until it finds fulfillment in the kingdom of God'" (Luke 22:14-16).

Passover is the annual Jewish celebration of the Exodus story, when God rescued his people out of slavery in Egypt and led them toward the Promised Land. This story, this festival, and this meal were and are essential to Jewish identity, both national and spiritual. But in the way Jesus shares this meal with his disciples, we see a reimagining and reorienting of the story: "And he took bread, gave thanks and broke it, and gave it to them, saying, 'This is my body given for you; do this in remembrance of me.' In the same way, after the supper he took the cup, saying, 'This cup is the new covenant in my blood, which is poured out for you'" (Luke 22:19-20).

In the original Passover Seder, there were three specific actions of blessing involved, all of which Jesus enacts by giving thanks, breaking the bread, and giving it to those around the table. But in this new meal, the bread and the cup take on new meaning. Now they represent Jesus' body and blood. And just as the original Seder was meant to remind the Jews that they were God's chosen people, who had been rescued and would be rescued again someday, this new meal identifies them as a particular people bound up together in a particular story of rescue; but Jesus expands the demarcating lines of who God's people are in revolutionary ways. Now, the people of God are not defined by ethnic, socioeconomic, political, or familial lines. The people of God are all those who eat of the bread and drink of the cup of Christ. And all are welcome at Christ's table. Jesus exemplified this reality, even long before the Last Supper.

Throughout the Gospels, Jesus is criticized for eating with the wrong people. This criticism was to be expected in the first-century Jewish world, as "the etiquette of the table [was] deeply significant in ordinary Jewish life, and textured with religious meaning. Among Jews in Jesus' day, who you ate with was as important as what you ate and how you ate."[3] Theologian Robert Karris even writes that, "Jesus was killed because of the way he ate."[4] While Karris might be overstating it a bit, his point is well taken. Jesus ate with people he had no business eating with. Or, better put, Jesus ate with people who had no business eating with him. And in doing so, even before his final meal, he redefined what it means to be the people of God around himself. He made a way for everyone—Jews and Gentiles—to belong.

This is why eating and drinking this meal still matters: because in doing so, we are eating and drinking our way back into an awareness of our place at God's table. Slowly but surely, we recognize that we are feasting with family, dining among a people to whom we eternally belong. I don't mean that this meal is some sort of golden ticket for salvation. But I also do not mean that it's just an optional, helpful practice. It's not. It's an undeniable and irresistible invitation to all those who belong to the body of Christ. In far too many of our churches, this meal has become a sporadic appetizer rather than the consistent base of our spiritual and communal nourishment. We forget that this meal was both the last and lasting thing Christ gifted to us before his death and resurrection. And in our forgetting, we risk losing our sense of what it means to be the people of God. Eating and drinking this meal together is what Christ gave us to remember who and whose we are.

For the vast majority of church history, followers of Jesus have gathered around the bread and the cup as the centerpiece of their worshiping community. As Andrew McGowan writes, the Jesus meal "was not a social event additional to worship, or a programmatic attempt to create fellowship among the Christians, but the regular form of Christian gatherings."[5] It was a meal that required the community to come together. It demanded, and continues to demand, that we gather *as* the body of Christ in order to *receive* the body of Christ as *one*. McGowan continues, "The body eaten is focused communally rather than individually, finding the Savior's presence in the corporate consumption rather than in the elements taken in isolation."[6] In our ever-so-isolated digital world, this meal offers us the opportunity to remind ourselves and the communities we serve that we are not alone, that Christ has come and is coming again, and that all those who eat of this bread and drink of this cup belong, to him and to one another, both here and now and forever.

THE MEAL AT THE CENTER OF HISTORY

In his letter to the Corinthians, Paul writes this regarding the meal Jesus gave us: "For whenever you eat this bread and drink this cup, you proclaim the Lord's death until he comes" (1 Corinthians 11:26). N. T. Wright points out that in Paul's explanation of what happens when we eat the bread and drink the cup, all three segments of time are represented: "The present moment ('whenever') somehow holds together the one-off past event ('the Lord's death') and the great future when God's world will be remade under Jesus' loving rule ('until he comes')."[7]

As technology speeds life up to unprecedented levels, people are becoming hurried and frenzied in unprecedented ways. Life is always hectic, and it's easy to lose the ability to see the long view of history unfolding, both behind and before us. But in the bread and the cup, all that God has been doing, is doing, and will eventually do is encapsulated into the basic human act of eating and drinking. It is the meal at the center of history. This is what we are inviting people into every time we pause to remember Christ through the bread and the cup. As Wright puts it, in this meal "God's past catches up with us again, and God's future comes to meet us once more."[8]

In the digital age, I can't think of a more important thing to do. I can't imagine why we wouldn't share this meal as often as we possibly can. I am greatly encouraged by the shift more and more churches seem to be making in recent years, placing Communion at the center of their worship gatherings. From non-denominational megachurches to denominational, high-liturgy churches, communities all over the country seem to be moving in this direction, shifting from sporadic "Communion Sundays" to inviting people to the table every time they gather to worship. This isn't a trend or a fad; it is a necessary return to the meal that has historically been the centerpiece of Christian worship. As we serve and lead our church communities, we must continue to guide them back to the table, to the bread and the cup, again and again. We must invite them to show up, hungry for the body of Christ as the body of Christ, because as much as modern technology wants to tell you so, you can't eat and drink together online.

BLINDED BY THE LIGHT

WHERE DO WE GO FROM HERE?

O N JANUARY 17, 1994, AT 4:30 A.M., a 6.7-magnitude earthquake struck Los Angeles, California. It was so intense that the ground shook as far away as Las Vegas, more than two hundred miles away from the epicenter. Electric lines were snapped, causing power outages as far north as Canada and as far east as Wyoming. In north Los Angeles, all power was lost. In the very early hours of that winter morning, about two million residents woke up to darkness.[1] People began to call 911.

That fact alone isn't at all surprising. But what they were calling about was. Numerous people reported an unidentifiable but clearly visible substance in the sky. Similar calls came in at various emergency centers as well as the Griffith Observatory. Residents anxiously described a "giant silvery cloud" ominously hanging over the city. People were afraid that an alien invasion was underway. They were wrong. What they were actually seeing was the Milky Way.

Los Angeles is often called the City of Light because of its luminosity. The sky glow emitted by the city's artificial light is visible from an airplane two hundred miles away.[2] In environments like Los Angeles, where light pollution is so severe, it's basically impossible to see the natural night sky lit up with stars.

Our artificial light is the very thing keeping us from seeing real light. In the words of Bruce Springsteen, we've been "blinded by the light."[3]

The digital age gleams and glows in magnificent ways. Like the brightest lights of our brightest cities, its technologies are tremendously beneficial when harnessed and leveraged wisely and responsibly. And yet, like the brightest lights of our brightest cities, its technologies also blind us to the sort of light we could never manufacture on our own. The kind of light we search for and wait for, even to catch just a glimpse, rather than the lights we easily turn on and off with a flick of a switch. As we consider what it means for us to serve and lead our churches well in the digital age, guiding them down more analog paths, we must remember that there's only one light that shines bright enough to lead us through the dimly lit glare of the digital age.

The subtitle of this book, "Why We Need Real People, Places, and Things in the Digital Age," is a bit of a misnomer. Maybe you thought it was questionable when you first read it. In the end, although analog realities are more important than ever, the only one we really need, the only one who can truly lead us where we long to go is Jesus Christ himself, who declared in John 8:12, "I am the light of the world. Whoever follows me will never walk in darkness, but will have the light of life."

When he spoke these words, Jesus was at the temple during a Jewish holiday known as Sukkot, the Feast of Tabernacles.

This eight-day fall festival commemorated a part of the Exodus story, when in the wilderness, God instructed the people of Israel to construct temporary shelters as a reminder of their constant readiness and willingness to move as God led them, as a pillar of cloud by day and fire by night. As a visual reminder of the story, the temple courts would have been illuminated by several tall flaming candelabras. David Brickner, executive director of Jews for Jesus, describes the scene this way: "Picture sixteen beautiful blazes leaping toward the sky from these golden lamps. Remember that the Temple was on a hill above the rest of the city, so the glorious glow was a sight for the entire city to see."[4] Against this backdrop of the bright lights, Jesus declares, "I am the light of the world" (John 8:12). Against the backdrop of all the bright lights of the digital age, Jesus is still declaring the same today.

So where do we go from here? We go where the Light leads us. Just like the Israelites in the wilderness, we learn to live and to lead in temporary shelters, at the ready to move when the Light moves, willing to change course, turn any direction, switch back, leap forward, hold still, and keep pace. We extinguish the urge to go our own way. We remember that although there are many bright lights *in* the world, there is only one Light *of* the world. And he's all the Light we need.

ACKNOWLEDGMENTS

J ENNY, THE WORDS "THANK YOU" seem far from enough. This book is your work in unique yet undeniable ways. Harper and Simon, I love you both more than you know.

There are so many others who have my deep and sincere gratitude. There aren't enough pages here to name them all but a few in particular . . .

Young Kim, my mother, for literally everything.

Dan Kimball, for believing in me and in this book from day one. Your friendship and example have changed my life and ministry.

Vintage Faith Church, for allowing these ideas to come to life in our community.

Steve Clifford, for your leadership, friendship, and investment in my life.

Isaac Serrano, for being the friend I text when I have a question about Aquinas. Collaborating with you is a joy.

Forrest Jenan, I'm glad we had all that free time at Sugar Pine way back.

Bryan Muirhead, I'm at Black Bear. Where are you?

Ethan McCarthy, for guiding me along and caring for this book with such care and precision.

Andrew Bronson, Helen Lee, and the entire team at IVP for believing in this work and all your effort releasing it into the wild.

Kristin Jensen, for lending your mind and your voice in shaping the conversation around these ideas; and for all those baby clothes.

Andy Gridley, David Kim, Dave Tieche, Josh Fox, for listening and laughing and letting me have the last crab rangoon.

Josh Butler, Chris Nye, Tim Mackie, Liz Ditty, and other friends who've let me ramble over the years about this book before it was ever a thing, for encouraging, prodding, and pushing me forward.

Andy Crouch, for your inspiration, feedback, and time. In many ways, you are this book's provocateur.

Scot McKnight, for the foreword; but more importantly, for your support over the years to the work and mission of the ReGeneration Project.

The countless friends, acquaintances, and strangers who've supported and continue to support the ReGeneration Project—your belief in the mission gives me immense hope.

WestGate Church, Awakening, Church on the Hill, you've each shaped me over the years and I wouldn't be here without you.

Pastors and church leaders who've been faithfully serving and humbly leading your communities toward Jesus Christ, with no fanfare and few applause, thank you. Your labor is not in vain.

DISCUSSION QUESTIONS

1. SLOW AND STEADY

1. What specific effects of the digital age (*speed/impatience, choices/shallowness, individualism/isolation*) do you see most affecting your church community right now? In what specific ways are you and your community experiencing these effects?

2. In what ways is your church a *derivative* of the cultural moment and in what ways is it a *disruption* to it? Are there specific things you might be able to do in order to shift the *derivative* elements toward becoming helpful *disruptive* elements to culture?

3. When it comes to the amount of time, energy, resources, and staffing, does your church focus more on "front door" experiences or "kitchen/living room" experiences? What, if anything, does this reveal about your priorities and potential shifts you may consider making?

4. Are there ways in which your church may be unintentionally leaning toward becoming a "digitally savvy business," under the guise of striving to become a "digitally savvy church"? Consider if there may be ways in which you're thinking of church in terms of *relevance* and *commodity*, rather than in terms of *transcendence* and *community*.

5. What are some specific, practical ways you can begin to help your church community live in "creative, prophetic" opposition to the perils of the digital age by "gathering when the world scatters, slowing down when the world speeds up, and communing when the world critiques"?

2. CAMERAS, COPYCATS, AND CARICATURES

1. How does your church community define and understand "worship"? How much does it align (or not align) with the sort of "whole-body participation" understanding of worship that's described in this chapter?

2. What people/churches/movements have most strongly influenced the philosophy, style, and culture of your church's worship gatherings? How and why did these people/churches/movements become your primary influences? Take some time to critically consider their influence. Are there specific ways your church may need to move in a different direction to better move forward into the future?

3. When it comes to the singing life of your church community, what sort of "gap" exists between those who lead the music up front and the rest of the gathered community? What specific and practical steps can you take to bridge this gap, in order to create a stronger invitation for people to be participants rather than spectators?

4. When it comes to the preaching and teaching life of your church community, do you and/or the leadership of your church view it primarily as "monologue" or

"dialogue"? What are some specific, practical ways to move to a more dialogical approach to preaching and teaching in your church?

5. Have you ever received input and feedback from younger generations that you're trying to reach regarding the way your church's worship gatherings are being experienced and perceived? If so, what does their feedback reveal? If not, what are some practical steps you can begin taking to receive such feedback consistently?

3. TO ENGAGE AND TO WITNESS

1. Think back to the first times you experienced God in a powerful, transformative way through music and preaching/teaching. What are some key elements that stand out to you? If working through this question with others, discuss the common themes that arise from everyone's experiences.

2. Consider the following equation from this chapter: *minimal resources + strong sense of community + awareness of need for Jesus = transcendent worship experience.*

3. What strikes you about this equation? Is anything surprising to you? Does it disrupt or validate any of your previously held assumptions about what matters most in cultivating spaces for people to worship God richly and deeply?

4. When it comes to the singing life of your church, where does your community land on the spectrum between *entertainment* and *engagement*? Are any potential changes or shifts necessary?

5. When it comes to the preaching/teaching life of your church, what sorts of markers or signs would indicate to you that people are truly *witnessing* the sermon and not simply *watching* it? In light of this, are any potential changes or shifts necessary?

6. How are you and those on your church leadership team building "a sense of rapport and trust . . . a sharing of life within the community"? What are some specific, practical ways to lean more in this direction at a leadership level?

7. Is there enough room and opportunity for joy and mourning, creativity and artistry within the worshiping life of your church? If so, how so? If not, why not?

4. REBUILDING BABEL

1. In what ways does your church encourage and equip people toward "patiently journeying" together and "doing the hard work of cultivating and excavating depth in our relationships with God and one another"? In what specific ways might your church do this even more effectively moving forward?

2. What are some specific, practical ways that your church can better become a "transcendent space where unlikely people gather to listen and speak, to reflect and respond, to journey together for the long haul down the path of wisdom"? Consider each element of that phrase and take time to unpack them, one by one.

3. Has your church community been affected by the "church shopping" phenomenon? In what specific ways has this mentality impacted the way you function as a church?

4. How does your church currently utilize online platforms to create a sense of community? In what ways are these platforms intentionally inviting people to participate in the actual, real-life community of the church? And in what ways might these platforms be communicating to people, intentionally or unintentionally, that church is a product to be consumed?

5. What are some of the best things your church is doing right now to cultivate real-life, in-person, transformative community amongst the people you serve?

5. A TAX COLLECTOR AND A ZEALOT WALK INTO A CROSSFIT

1. What are your thoughts on the idea that "churches, at their best, bring us into contact with people we would never think of as friends," that necessarily and by design the church forms the "unlikes" into a fellowship? Is this a reality you see being lived out in your church? Why or why not?

2. In your church, is there a culture of working through the "mess, complications, and inconveniences" of real relationships and community together? How so?

3. In what specific, practical way does your church dedicate the time, energy, and resources to being present with people in pain?

4. Who in your church (both staff and lay leaders) is equipped and empowered to do the important work of consistent pastoral care to your community?

5. How are you teaching, inspiring, and equipping your church to do the difficult yet necessary work of learning and living out the art of confession?

6. Does your church ever "stay and feast" together? How regularly? What are some practical ways you can create more opportunities for this?

6. JACKPOT!

1. How have you engaged the Bible throughout your life? Track the different stages of your journey/relationship with the Bible.

2. Does understanding the early church's communal and long-form engagement with the Bible change your perceptions of how you ought to engage in any way? Or, does it affirm your current approach?

3. Nicholas Carr writes, "Once I was a scuba diver in a sea of words. Now I zip along the surface like a guy on a Jet Ski." If you had to guess, how does the majority of your church community engage the Bible—are they "scuba divers" or "jet skiers"?

4. How are you reading and teaching the Bible in a way that it both *comforts* and *confronts* in your church?

5. What are specific, practical ways you can begin to equip and encourage your church community to read and engage the Bible deeply, patiently, and together—to "swim in the deep end," so to speak?

7. HOWTOREADABOOK

1. Think back to a time when you experienced God speak to you in a powerful and transformative way through the Bible. What do you remember about the experience?

2. Consider Adler and Van Doren's four key questions from *How to Read a Book*:

   - What is the book about as a whole?
   - What is being said in detail, and how?
   - Is the book true, in whole or part?
   - What of it?

3. Which of these questions do you most consistently consider when reading the Bible? Which do you often neglect?

4. What is the overall approach at your church regarding the style of sermons? Is it usually felt-need series or going through longer passages or entire books of the Bible? Does the reality that most people "read for the jackpot" shift how you might format your preaching/teaching going forward?

5. What changes to your approach to sermon series may be necessary in order to better tell the whole story of the Bible as consistently as possible?

6. How have you previously defined what it means to be a "theologian"? How does embracing the truth that "we are all theologians" change the way you think about equipping and encouraging your church to read and exegete the Bible?

7. What are some specific, practical ways to help move your church community from reading for *information*

(falling into the trap of theological elitism where having the right answer is the goal) to reading for *transformation* (becoming the right person in light of who Christ is, what he's done and what he's calling us to become)?

8. THE MEAL AT THE CENTER OF HISTORY

1. Are there any ways you've been caught up in the digital age's preference for efficiency and convenience, specifically in regards to how you serve and lead your church?

2. In your church, is Communion "a sporadic appetizer" or is it a "consistent base of spiritual and communal nourishment"? How often do you share the bread and the cup together? If not every time you gather to worship, why not? What's the philosophy behind the rhythm for Communion?

3. Andrew McGowan writes, "The body eaten is focused communally rather than individually, finding the Savior's presence in the corporate consumption rather than in the elements taken in isolation." Do you agree or disagree? How might thinking of it as a communal act change the way communion is taught and experienced in your church?

NOTES

INTRODUCTION: EDM AND GRANDMA'S CHURCH

[1]Chris Nye, *Less of More* (Ada, MI: Baker Books, 2019), 70.

[2]Sherry Turkle, *Alone Together*, rev. and exp. ed. (New York: Basic Books, 2017), 1.

[3]This idea, "The medium is the message," is one of the primary thrusts in Marshall McLuhan, *Understanding Media* (Corte Madera, CA: Gingko Press, 2003).

1. SLOW AND STEADY

[1]Dallas Willard, *The Great Omission* (San Francisco: HarperOne, 2014), xv.

[2]Kevin McSpadden, "You Now Have a Shorter Attention Span Than a Goldfish," *Time*, May 14, 2015, http://time.com/3858309 /attention-spans-goldfish.

[3]Simon Maybin, "Busting the Attention Span Myth," BBC News, May 10, 2017, www.bbc.com/news/health-38896790.

[4]*The Shallows* by Nicholas Carr and *Irresistible* by Adam Alter, both of which I cite later in the book, are thought provoking and enlightening resources into this research.

[5]One of the most helpful books in recent memory when it comes to figuring out how to apply wisdom and character in our approach to digital technology is Andy Crouch, *The Tech-Wise Family* (Ada, MI: Baker Books, 2017).

[6]Crouch, *The Tech-Wise Family*, 66.

[7]Adam Alter, *Irresistible* (New York: Penguin Books), 28.

[8]"Internet Stats," Omnicore Agency, www.omnicoreagency.com /category/internet-stats.

[9]Cal Newport, *Deep Work* (New York: Grand Central Publishing, 2016), 7.

[10]Newport, *Deep Work*, 160.

[11]One of the best works I've read on how our desires are at the core of both who we are and who we're becoming is James K. A. Smith, *You Are What You Love* (Ada, MI: Brazos Press, 2016).

[12]C. S. Lewis, *Weight of Glory*, rev. ed. (San Francisco: HarperOne, 2001), 26.

[13]Sherry Turkle, *Alone Together*, rev. and exp. ed. (New York: Basic Books, 2017), 154.

[14]Willard, *Great Omission*, 25.

[15]James F. White, *The Oxford History of Christian Worship*, ed. Geoffrey Wainwright and Karen B. Westerfield Tucker (New York: Oxford University Press, 2006), 804.

[16]Alan Noble, *Disruptive Witness* (Downers Grove, IL: InterVarsity Press, 2018), 120.

[17]Willard, *The Great Omission*, 65.

[18]David Sax, *The Revenge of Analog* (New York: PublicAffairs, 2016), 10.

[19]Sax, *Revenge of Analog*, 8.

[20]Paul Taylor, "Amazon to Open 3,000 Brick-and-Mortar Stores by 2021," *Techspot*, September 19, 2018, www.techspot.com /news/76527-amazon-open-3000-brick-mortar-stores-2021.html.

2. CAMERAS, COPYCATS, AND CARICATURES

[1]Daniel Block, *For the Glory of God* (Ada, MI: Baker Academic, 2016), 8.

[2]See Romans 7:4; 1 Corinthians 10:16-17, 12:12-27; Ephesians 1:22-23, 4:16, 5:29-32; and Colossians 1:18, 1:24, 3:15.

[3]Walo Fenton, "CBC TV—Take 30 (Program)—McLuhan Predicts 'Word Connectivity' 1965," YouTube, March 30, 2010, www .youtube.com/watch?v=NNhRCRAL6sY.

[4]Marshall McLuhan, *Understanding Media* (Corte Madera, CA: Gingko Press, 2003), 70-71.

[5]"Charles Wesley's Final Hymns," The United Methodist Church Discipleship Ministries, www.umcdiscipleship.org/resources /charles-wesleys-final-hymns.

[6]Tim Challies, "What We Lost When We Lost Our Hymnals," Challies, March 29, 2017, www.challies.com/articles/what-we -lost-when-we-lost-hymnals.

[7]Alan Noble, *Disruptive Witness* (Downers Grove, IL: InterVarsity Press, 2018), 122.

[8]David Fitch, "The Two Sins of Multisite Video Venue," Missio Alliance, December 11, 2014, www.missioalliance.org/the-two-sins

-of-multi-site-video-venue-the-case-of-mark-driscollmars-hills
-and-how-should-we-respond.

9Thomas Long, *The Witness of Preaching*, 2nd ed. (Louisville, KY: Westminster John Knox, 2005), 26.

10Jake Meador, "Should Sermons Be Published?" Mere Orthodoxy, August 21, 2014, https://mereorthodoxy.com/should-sermons -be-published.

3. TO ENGAGE AND TO WITNESS

1Dietrich Bonhoeffer, *Life Together: Prayerbook of the Bible* (Minneapolis: Fortress, 1995), 66.

2Andy Crouch, *The Tech-Wise Family* (Ada, MI: Baker Books, 2017), 184-85.

3Manuel Luz, *Honest Worship* (Downers Grove, IL: InterVarsity Press, 2018), 14-15.

4Scott Johncock, "Doxology at New Life Downtown," YouTube, March 27, 2016, www.youtube.com/watch?v=c_raZpY0XsI.

5Anna Keating, "Why Evangelical Megachurches Are Embracing (Some) Catholic Traditions," *America, The Jesuit Review*, May 2, 2019, www.americamagazine.org/faith/2019/05/02/why-evangelical -megachurches-are-embracing-some-catholic-traditions.

6Crouch, *Tech-Wise Family*, 186.

7The "easy-everywhere" description of digital technologies is from Crouch, *Tech-Wise Family*.

8Jacques Launay and Eiluned Pearce, "The New Science of Singing Together," *Greater Good Magazine*, December 4, 2015, https:// greatergood.berkeley.edu/article/item/science_of_singing.

9Bonhoeffer, *Life Together*, 68.

10Thomas Long, *The Witness of Preaching*, 2nd ed. (Louisville, KY: Westminster John Knox, 2005), 16.

11Timothy Keller, *Preaching* (New York: Viking, 2015), 17-18.

12Christo and Jeanne-Claude, "The Gates," *Christo and Jeanne Claude*, https://christojeanneclaude.net/projects/the-gates.

13Michael Kimmelman, "A Billowy Gift to the City, in a Saffron Ribbon," *New York Times*, February 13, 2005, www.nytimes.com /2005/02/13/nyregion/a-billowy-gift-to-the-city-in-a-saffron -ribbon.html.

[14]Leonard Sweet, *Giving Blood* (Grand Rapids: Zondervan, 2014), 22.

[15]"The Gospel Going Forth in DFW and Beyond," The Village Church, http://multiply.thevillagechurch.net.

[16]"February 26 Announcement," Redeemer Church and Ministries, https://www.redeemer.com/r/february_26_announcement.

[17]Based on the immensely popular PreachersNSneakers Instagram account, I've quite literally never been in their shoes.

[18]Long, *Witness of Preaching*, 226.

[19]James K. Wellman Jr., Katie E. Corcoran, Kate Stockly-Meyerdirk, "'God Is Like a Drug . . .': Explaining Interaction Ritual Chains in American Megachurches," Wiley Online Library, August 26, 2014, https://onlinelibrary.wiley.com/doi/abs/10.1111/socf.12108.

4. REBUILDING BABEL

[1]Sherry Turkle, *Alone Together*, rev. and exp. ed. (New York: Basic Books, 2017), 267.

[2]For a succinct, easy-to-read description of ziggurat structures in the ancient world and the connection to the Genesis 11 story, check out this blog post from scholar John Walton: https://zondervanacademic.com/blog/tower-of-babel.

[3]"Why Americans Go (and Don't Go) to Religious Services," Pew Research Center, August 1, 2018, www.pewforum.org/2018/08/01/why-americans-go-to-religious-services.

[4]"Church Attendance Trends Around the Country," Barna Research Group, May 26, 2017, www.barna.com/research/church-attendance-trends-around-country.

[5]MrTechinformation, "Facebook Home 'Dinner'—Interesting Family," YouTube, April 16, 2013, www.youtube.com/watch?v=yF3Nk4YlU_Y.

[6]Dietrich Bonhoeffer, *Life Together: Prayerbook of the Bible* (Minneapolis: Fortress, 1995), 36.

[7]Sherry Turkle, *Reclaiming Conversation* (New York: Penguin Books, 2016), 154.

[8]Turkle, *Reclaiming Conversation*, 154-56.

[9]Turkle, *Reclaiming Conversation*, 323.

[10]James Emery White, *Meet Generation Z* (Ada, MI: Baker Books, 2017), 44.

[11]Church Platform Online, part of the Life.Church Open Network, https://churchonlineplatform.com.

[12]Life.Church Online, "What Is Church Online?," YouTube, October 18, 2016, www.youtube.com/watch?v=yPKNEmQMO0A.

[13]Caroline Golum, "8 Must-Know Statistics About Livestreaming for Houses of Worship," Livestream, https://livestream.com/blog/livestreaming-houses-worship-statistics.

[14]See Amish America at http://amishamerica.com.

[15]TJ Barnes, quoted in Golum, "8 Must-Know Statistics About Livestreaming."

[16]Turkle, *Alone Together*, 227.

[17]Judah Smith, Twitter post, Nov. 4, 2018, https://twitter.com/judahsmith/status/1059194945100107776.

[18]Ed Stetzer, "Is Online Church Really a Church?" *Christianity Today*, April 2, 2014, www.christianitytoday.com/edstetzer/2014/april/is-online-church-really-church.html.

5. A TAX COLLECTOR AND A ZEALOT
WALK INTO A CROSSFIT

[1]John Shinal, "Mark Zuckerberg: Facebook Can Play a Role that Churches and Little League Once Filled," CNBC, June 26, 2017, www.cnbc.com/2017/06/26/mark-zuckerberg-compares-facebook-to-church-little-league.html.

[2]Felix Allen, "Mark Zuckerberg Says Facebook Is 'the New Church,'" *New York Post*, June 29, 2017, https://nypost.com/2017/06/29/mark-zuckerberg-says-facebook-is-the-new-church.

[3]Peter Ormerod, "Mark Zuckerberg, the Church of Facebook Can Never Be. Here's Why," *The Guardian*, June 29, 2017, www.theguardian.com/commentisfree/2017/jun/29/mark-zuckerberg-church-facebook-social-network.

[4]Scot McKnight, *Fellowship of Differents* (Grand Rapids: Zondervan, 2015), 27.

[5]Joseph Hellerman, *When the Church Was a Family* (Nashville: B&H Academic, 2009), 124.

[6]For an in-depth analysis, see Hellerman, *When the Church Was a Family*, 34-52.

[7]See 1 Corinthians 6:1-8.

[8]Brett McCracken, *Uncomfortable* (Wheaton, IL: Crossway), 34.

[9]Jaron Lanier, *Ten Arguments For Deleting Your Social Media Accounts Right Now* (New York: Henry Holt and Co., 2018), 74.

[10]Dietrich Bonhoeffer, *Life Together: Prayerbook of the Bible* (Minneapolis: Fortress, 1995), 34.

[11]Christine Wang, "How a Health Nut Created the World's Biggest Fitness Trend," CNBC, April 5, 2016, www.cnbc.com/2016/04/05 /how-crossfit-rode-a-single-issue-to-world-fitness-domination.html.

[12]Bonhoeffer, *Life Together*, 96.

[13]Sebastian Junger, *Tribe* (New York: Twelve, 2016), 66.

[14]Bonhoeffer, *Life Together*, 110.

[15]"Social Eating Connects Communities," University of Oxford, March 16, 2017, www.ox.ac.uk/news/2017-03-16-social-eating -connects-communities.

[16]The Abbey in Santa Cruz, CA; Red Rock Coffee in Mountain View, CA; Café 4 in Castro Valley, CA; and Community Coffee in Milpitas, CA.

6 JACKPOT!

[1]See the Center for Humane Technology at http://humanetech.com.

[2]Tristan Harris, "How Technology Is Hijacking Your Mind—from a Magician and Google Design Ethicist," *Thrive Global,* May 18, 2016, https://medium.com/thrive-global/how-technology-hijacks -peoples-minds-from-a-magician-and-google-s-design-ethicist -56d62ef5edf3.

[3]Alexis C. Madrigal, "The Machine Zone: This Is Where You Go When You Just Can't Stop Looking at Pictures on Facebook," *Atlantic,* July 31, 2013, www.theatlantic.com/technology/archive /2013/07/the-machine-zone-this-is-where-you-go-when-you -just-cant-stop-looking-at-pictures-on-facebook/278185.

[4]Nicholas Carr, *The Shallows* (New York: W. W. Norton & Company, 2011), 117.

[5]Andrew Perrin, "Who Doesn't Read Books in America?," Pew Research Center, March 23, 2018, www.pewresearch.org/fact -tank/2018/03/23/who-doesnt-read-books-in-america.

[6]Thomas Pettitt, "Opening the Gutenberg Parenthesis: Media in Transition in Shakespeare's England," *Academia,* August 4, 2013, www.academia.edu/2946207/Before_the_Gutenberg _Parenthesis_Elizabethan-American_Compatibilities.

[7]Carr, *Shallows*, 127.

[8]Carr, *Shallows*, 63.

[9]Larry Hurtado, *Destroyer of the Gods* (Waco, TX: Baylor University Press, 2016), 108.

[10]For a thorough analysis of this, read Brian J. Wright, *Communal Reading in the Time of Jesus* (Minneapolis: Fortress, 2017).

[11]Carr, *Shallows*, 6-7.

[12]She Reads Truth, "Use today to pray, rest, and reflect on this week's reading, giving thanks for the grace that is ours in Christ," Instagram, November 10, 2018, www.instagram.com/p /BqAKDONBUI8.

[13]Kevin Vanhoozer, *Biblical Authority After Babel* (Ada, MI: Brazos Press, 2018), 88.

[14]Cal Newport, *Deep Work* (New York: Grand Central Publishing, 2016), 84.

[15]Carr, *Shallows*, 138.

[16]Alan Jacobs, *The Pleasures of Reading in an Age of Distraction* (Cambridge, MA: Oxford University Press, 2011), 16.

[17]The Bible Project is further referenced in the next chapter. There may not be a better, more accessible online resource to aid in Bible reading and study than this: https://thebibleproject.com.

[18]Richard Dawkins, *The God Delusion* (Boston: Mariner Books, 2008), 51.

7. HOW TO READ A BOOK

[1]Mark Seidenberg, "Sorry, But Speed Reading Won't Help You Read More," *Wired,* January 24, 2017, www.wired.com/2017/01 /make-resolution-read-speed-reading-wont-help.

[2]Seidenberg, "Sorry."

[3]Paul Saenger, *Space Between Words* (Redwood City, CA: Stanford University Press, 2000), 11.

[4]Two words in Genesis 31:47 plus Ezra 4:8–6:18, 7:11-26; Jeremiah 10:11; and Daniel 2:4–7:28 are written in Aramaic.

[5]Nicholas Carr, *The Shallows* (New York: W. W. Norton & Company, 2011), 61.

[6]Carr, *Shallows*, 50-51.

[7]Mortimer Adler and Charles Van Doren, *How To Read a Book,* rev. ed. (New York: Touchstone, 1972), 46-47.

[8]Adler and Van Doren, *How To Read a Book*, 47.

[9]Adler and Van Doren, *How To Read a Book*, 47.

[10]Adler and Van Doren, *How To Read a Book*, 47.

[11]Dietrich Bonhoeffer, *Life Together: Prayerbook of the Bible* (Minneapolis: Fortress, 1995), 60.

[12]Scot McKnight, *The Blue Parakeet* (Grand Rapids: Zondervan, 2008), 44.

[13]See https://thebibleproject.com for more information.

[14]See www.youtube.com/user/jointhebibleproject/videos to view their videos.

[15]The Bible Project, "BTS 7: The Revelation Premiere," YouTube, December 1, 2016, www.youtube.com/watch?v=1YD__DCKb7o&t =4093s.

[16]A. W. Tozer, *Knowledge of the Holy* (New York: HarperCollins Publishers, 1961), 1.

[17]For those interested in further reading on biblical exegesis, there are many resources out there. One place to start that I'd suggest is Michael J. Gorman, *Elements of Biblical Exegesis*, rev. and exp. ed. (Ada, MI: Baker Academic, 2010).

[18]Gordon Fee and Douglas Stuart, *How to Read the Bible for All It's Worth*, 3rd ed. (Grand Rapids: Zondervan, 2003), 24.

[19]Find out more about the Year of Biblical Literacy at http://bible .realitysf.com.

[20]Learn about the ReGeneration Project at www.regeneration project.org.

[21]Lesslie Newbigin, *A Walk Through the Bible* (Vancouver: Regent College Publishing, 2005), 13.

[22]McKnight, *Blue Parakeet*, 52.

[23]Dallas Willard, *The Great Omission* (San Francisco: HarperOne, 2014), 52.

[24]"The Bible in America: 6-Year Trends," Barna Research Group, June 15, 2016, www.barna.com/research/the-bible-in-america -6-year-trends.

[25]Sherry Turkle, *Reclaiming Conversation* (New York: Penguin Books, 2016), 13.

8. THE MEAL AT THE CENTER OF HISTORY

[1]Lee Vander Zee, *Christ, Baptism, and the Lord's Supper* (Downers Grove, IL: IVP Academic, 2004), 192.

[2]For a helpful, concise explanation of how these various names came to represent the one meal, see N. T. Wright, *The Meal Jesus Gave Us*, rev. ed. (Louisville, KY: Westminster John Knox, 2015), 37-41.

[3]Vander Zee, *Christ, Baptism, and the Lord's Supper*, 141.

[4]Robert Karris, *Luke: Artist and Theologian* (Mahwah, NJ: Paulist Press, 1985), 70.

[5]Andrew McGowan, *Ancient Christian Worship* (Ada, MI: Baker Academic, 2016), 19.

[6]McGowan, *Ancient Christian Worship*, 32.

[7]Wright, *The Meal Jesus Gave Us*, 51.

[8]Wright, *The Meal Jesus Gave Us*, 50.

CONCLUSION: BLINDED BY THE LIGHT

[1]Anthony Ramirez, "The Earthquake: Phones and Power; Gas and Electric Services Are Disrupted for Millions of Customers," *New York Times*, January 18, 1994, www.nytimes.com/1994/01/18/us/earthquake-phones-power-gas-electric-services-are-disrupted-for-millions.html.

[2]Ron Chepesiuk, "Missing the Dark: Health Effects of Light Pollution," *Environmental Health Perspectives* 117, no. 1 (January 2009): 20-27, www.ncbi.nlm.nih.gov/pmc/articles/PMC2627884.

[3]Bruce Springsteen, "Blinded by the Light," *Greetings from Asbury Park, N.J.*, Columbia Records, 1973.

[4]David Brickner, "Finding Jesus in the Feast of Tabernacles," The Christian Broadcasting Network, www1.cbn.com/finding-jesus-feast-tabernacles.

ReGENERATION
PROJECT

The ReGeneration Project is a collective of church leaders, theologians, artists, college students, and young adults who are passionate about helping new generations follow Jesus and join him on mission in the world.

Through events, podcasts, books, articles, and other resources, the ReGeneration Project exists to encourage and equip teens, college students, young adults, and those passionate about serving and reaching them to engage the Bible, theology, and the mission of the church in new ways.

CONNECT WITH JAY AT

jaykimthinks.com

and @jaykimthinks (Twitter and Instagram)